Plastics Materials: Properties and Applications

Plastics Materials
Properties and Applications
Second Edition

A.W. BIRLEY
R.J. HEATH
Institute of Polymer Technology and Materials Engineering
University of Technology
Loughborough, UK

M.J. SCOTT
Lowe & Fletcher Ltd
Telford, UK

Blackie

Glasgow and London

Published in the USA by
Chapman and Hall
New York

Blackie & Son Limited,
Bishopbriggs, Glasgow G64 2NZ
7 Leicester Place, London WC2H 7BP

Published in the USA by
Chapman and Hall
in association with Routledge, Chapman and Hall, Inc.
29 West 35th Street, New York, NY 10001–2291
© 1988 Blackie & Son Ltd
First published 1982
This edition 1988

British Library Cataloguing in Publication Data

Birley, Arthur W.
    Plastics materials: properties and
    applications.—2nd ed.
    1. Plastics
    I. Title   II. Heath, Richard   III. Scott,
    Martyn J.
    668.4

    ISBN 0–216–92489–8
    ISBN 0–216–92490–1 Pbk

For the United States, International Standard Book Number is

0–412–01771–7
0–412–01781–4 (pbk)

Phototypesetting by Thomson Press (India) Ltd., New Delhi
Printed in Great Britain by Bell & Bain (Glasgow) Ltd

Preface

Plastics are part of everyday life and contribute immensely to the benefit of humanity. When failures occur, they are due in part either to inferior properties (resulting from poor design or badly controlled processing), or to an incomplete understanding of the properties and applications of plastics materials.

Since publication of the first edition, the plastics industry has increasingly adopted advanced business procedures and automation (such as closed loop control and robotics), to combat the effects of recession and has moved increasingly towards methods based on sound scientific and technological principles. Plastics have increasingly been used in applications once dominated by the use of metals and ceramics. For instance, in the automotive industry, the modern car now contains a much higher proportion of polymers, including commodity plastics and more specialized materials. In addition, compact discs are being made from new injection-moulding grades of polycarbonate, which meet the requirements of a demanding process.

This second edition has been thoroughly revised and extended to include new materials, technologies and design concepts. Chapters on thermoplastics reflect the development of polymer blends and alloys, whilst the chapters devoted to thermosets have been reorganized to accommodate the renaissance in the applications of phenolics and to cover the growing importance of polyurethanes. The related two-component process technologies are now included; having undergone major developments in the last decade, they have become important shaping processes.

We hope that readers, whether new to or experienced in plastics technology, will find in this book something that will extend their knowledge and increase their understanding of the field.

AWB
RJH
MS

Contents

1 Introduction

1.1 Background

Many plastics are now used in such quantities that they have reached the status of commodity materials; indeed, the volume usage of plastics now comfortably exceeds that of metals. Much of the growth has taken place over the last thirty-five years, and the plastics industry is still expanding at twice the rate of the economy as a whole. The motivation for the rapid growth is the suitability of plastics for mass production, which depends mainly on their easy and reproducible shaping, and their low volume cost, coupled with some attractive properties. Shaping at low temperatures into complex forms is a characteristic of most plastics, and ensures their increasing use in spite of some shortcomings. The manufacturing industry has responded very positively to the increasing demand for plastics and for diversification of properties. The major feedstock is oil; the dependent petrochemicals industry supplies the monomers for plastics production, and manufactures a wide range of additives to modify their behaviour.

The principal structural feature of most plastics is that a unit, or *mer*, is linked chemically many times (*poly*). The chemical process for the formation of such materials is known as *polymerization* and the products are *polymers*. Plastics cannot be defined precisely, although they are undoubtedly polymers and, by dictionary definition, they must be capable of being shaped at some stage in their history. Plastics are generally rigid, in contradistinction to rubbers (elastomers), which are very flexible.

1.2 Structure and properties of plastics

There are two subgroups of plastics: thermoplastics and cross-linked plastics. The former are linear chain polymers which soften on heating and solidify again on cooling, whereas the latter are network structures (in three dimensions) which, once formed, are not softened by heating. The repetition of units in a polymer can result in a regularity of structure which may have important consequences; in favourable cases a number of chains may become aligned and in register for some distance, a state which is favoured on energy grounds, and which is termed *crystallinity*. The occurrence of crystallinity in a polymer affects profoundly both processing and properties, and will be referred to frequently in the later chapters of this book.

'Properties' provide quantitative information on the response of a material to external stimuli, which can be mechanical, thermal, electrical, optical, chemical, etc., or combinations of these. Thus, if a material is stressed mechanically, it responds by deforming or breaking, the relevant properties being 'modulus' or 'strength'. Application of thermal energy to a sample causes it to increase in temperature, the relationship being termed the 'enthalpy' or 'heat content', and to increase its dimensions ('thermal expansion'). Consideration of the properties of plastics allows a first-order distinction to be made between those properties which depend on the *organic nature* (involving the element carbon), and which are based on short-range interactions, particularly the repeat unit, and those which result from the long-chain or network structure. The former include density, dielectric permittivity, melting and softening temperatures, and the interaction with solvents, acids and bases. The polymeric nature leads to properties which are unique, exemplified by the very high viscosities of solutions and melts of linear polymers, and the accommodation of very high deformation without fracture, with the possibility of almost complete recovery on removing the stress. In view of the great importance of the chain length or molecular weight of a polymer, many methods have been developed for its determination; quality control by molecular-weight (MW) measurement is now not uncommon. Monitoring MW-dependent properties, such as melt viscosity, is still practised widely.

In a more detailed consideration of properties, there is interaction between those dependent on short-range interactions and, in particular, the nature of the repeat unit in the polymer and the constraints imposed by the chain or network structure. For example, the density of a plastic is that expected of an organic substance (0.8 to 1.4 $Mg\,m^{-3}$). On cooling, the material changes density continuously as it strives towards equilibrium, but is constrained by the polymer structure. This *time dependence* is characteristic of polymers, particularly thermoplastics, and is a factor of considerable importance in their technology.

Properties are determined for many reasons, ranging from assurance of the quality of a product to providing the basis for component design and predicting its service life. These different requirements lead to different test philosophies, for quality control is concerned with the reproducibility of a simple, cheap but discriminating test procedure for fixed (but arbitrary) test parameters, while data for design are required over wide ranges of important variables, and are thus much more expensive to obtain. This book will be mainly concerned with design data and their application: an important source of information is BS 4618, Recommendations for the Presentation of Design Data for Plastics (British Standards Institution, London).

1.2.1 *Mechanical properties*

Adequate mechanical properties are a prerequisite in most applications of plastics; response to stressing by fracture or deformation is usually of crucial importance, while friction, wear, vibration damping etc., may be relevant in particular applications. For metals, it is commonly accepted that the behaviour can be forecast by stress analysis, thence using the modulus or design stress in the appropriate formula. A similar procedure has been shown to be applicable to plastics for stiffness in closely controlled experiments on carefully selected structural elements which are homogeneous and isotropic, provided that the appropriate value of 'modulus' is employed. This will be exemplified and discussed in the next chapter.

Breakage of a component or article is frequently a limitation on its use; this subject has, therefore, received considerable attention. Failure may be brittle or ductile, the former occurring at low elongation, the latter with considerable deformation, and as the culmination of creep. Brittle failure in plastics materials is sometimes inherent and sometimes unexpected. Many laboratory tests indicate that most plastics are either tough or brittle, which is very convenient, since there are rules for dealing with ductile materials, and different rules for dealing with brittle ones. However, some plastics break at unexpectedly low stresses and, even more seriously, products made from plastics which are expected to be ductile from laboratory tests and general experience can also fail in a brittle manner.

Considerations based on a *fracture mechanics* analysis are necessary if the designer is to progress beyond the primary division of plastics into 'ductile' or 'brittle', based on simple laboratory tests (see Chapter 2).

1.2.2 *Thermal properties*

The thermal properties of plastics are generally those expected for organic substances, modified in some cases by the polymer structure. Thus, partially

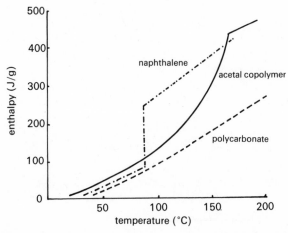

Figure 1.1 Enthalpy (heat content) vs. temperature. Comparison of data for typical crystalline material (naphthalene) with those for crystalline (acetal copolymer) and amorphous (poly-carbonate) plastics

crystalline plastics have melting points, although the fusion process is gradual. Thermal conductivity is low; conversely, thermal expansion and contraction are high. The relationship between heat energy applied and the resulting rise in temperature, the *enthalpy* or *heat content*, is generally as expected.

Thermal properties are relevant both to processing and to behaviour in the solid state. For partially crystalline plastics, the melting temperature defines the upper limit of form stability, although the plastic may soften unacceptably at lower temperatures. In processing, the enthalpy defines the quantity of heat which is necessary to change the temperature from ambient to the processing temperature. For crystalline plastics this will include a substantial contribution from the latent heat of fusion of the crystalline phase; indeed, the heat content at the processing temperature may well be a factor of two higher for a crystalline plastic compared with an amorphous one. It must be remembered that this heat has to be supplied during the heating stage and, more importantly, removed during cooling. This is illustrated in Figure 1.1, where data for a partially crystalline plastic, (acetal copolymer), are compared with those for an amorphous polymer (polycarbonate), and a crystalline low-molecular-weight organic material (naphthalene).

Differences between crystalline and amorphous plastics can be found also in thermal expansion and contraction, where the generally enhanced density of the crystalline regions results in greater differences between melt- and solid-phase densities, and aggravation of the mould shrinkage problem, (Figure 1.2). The heating and cooling of plastics masses is retarded by their low thermal conductivities, causing these operations to be time-dominating in many processes; for example, thermal conductivity of copper is $385\,\mathrm{W\,m^{-1}\,K^{-1}}$, that of polypropylene, $0.2\,\mathrm{W\,m^{-1}\,K^{-1}}$, whereas cellular polymers have excellent insulation properties, $0.02\,\mathrm{W\,m^{-1}\,K^{-1}}$ for rigid polyurethane foam.

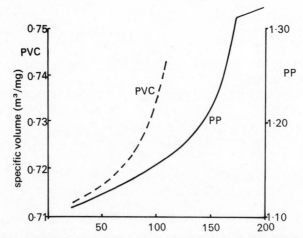

Figure 1.2 Thermal expansion of crystalline (polypropylene) and amorphous (poly(vinyl-chloride)) plastics

1.2.3 *Electrical properties*

As would be expected from their structures, most plastics, at least in the solid state, are very poor conductors of electricity; indeed, some are amongst the best insulators known. Dielectric properties—relative permittivity and power factor (or loss factor)—are typical of those found in low-molecular-weight organic materials, although the fine structure of such properties is affected by features associated with the structure. Conversely, the measurement of dielectric properties, which can be made with high precision over very wide frequency ranges, provides a powerful means of probing polymer structure.

The insulation afforded by plastics materials with the poorest properties, such as polyamide 6.6 or plasticized PVC, is still more than adequate, except for the most critical uses. This is exemplified by the use of the former by British Rail as an insulant in automatic track signalling, and of the latter in domestic cable insulation.

The intrinsic electrical strength of most plastics is higher by a factor of a thousand than the breakdown strength of air, so that in most practical situations breakdown is dominated by the breakdown of the air.

1.2.4 *Optical properties*

While the refractive properties are not unusual, covering a refractive index range of 1.35–1.65 for solid (unfoamed) materials, the transparency can vary considerably: some plastics are almost perfectly transparent, and find application as optical components, whereas others are translucent or opaque. The difference in transparency depends on the microstructure; scattering of light occurs from multiphase structures, the phases differing in refractive index and being of a size comparable with the wavelength of light. Scattering of light and a consequent deterioration in transparency can also result from surface irregularities, again on the scale of the wavelength of light, which can be introduced during processing. Such surface structure is frequently associated with *haziness* in polyethylene packaging film. A further aspect of surface scattering is a decrease in the *gloss*.

Colour is another aspect of optical properties, but here, plastics do not introduce any new phenomena, rather it is the behaviour of the colourants which is of interest: they must be stable during processing and use, and should not induce any reaction, either physical or chemical, in the polymer. It has been observed that for partially crystalline polymers, colourants frequently behave as nucleating agents.

1.2.5 *Melt properties*

As noted previously, the widespread use of plastics is based on the ease with which they can be shaped. Thermoplastics are rendered tractable by the application of heat, the shaped part being stabilized in form by cooling; no chemistry is involved. Cross-linking plastics, on the other hand, are shaped as comparatively low-molecular-weight prepolymers, stabilization of shape being achieved by cross-linking by specific chemical reactions. For both classes of plastics, the ease of shaping is related to the fluidity of the material: the mobility of prepolymers and the high viscosity of thermoplastics both limit the range of processing techniques which can be applied. Processing will be treated in greater detail later in this chapter.

1.2.6 *Chemical properties*

These may be subdivided into physical phenomena and those where there is chemical reaction: the former include the transmission of fluids through plastics barriers (permeation), and the interaction of plastics with solvents, the effects ranging from swelling to solution. Chemical reaction causes permanent change to the polymer.

The low density of plastics implies a relatively open structure which can be penetrated by fluids, such as water, oxygen or carbon dioxide, during which the barrier material may be affected only marginally. Permeation is especially important in packaging.

The interaction of plastics with solvents is relevant to the wider use of plastics materials. Solution (for linear polymers) or maximum swelling (for cross-linked systems) is favoured by similarity between polymer repeat unit and the solvent, and by specific interaction between solvent and polymer. On the other hand, solubility is reduced by crystallinity in the plastic, the energy associated with the formation of the crystallites having to be overcome before solution can be effected. Thus, crystalline plastics are considerably more resistant to solvents than are amorphous materials.

Although plastics are not particularly reactive chemically, and some are distinctly inert, the extent of chemical changes in plastics is useful information for the designer. An important class of reactions is that associated with *ageing*, which, as the term implies, is deterioration in performance which is progressive with time. Further, some environments are aggressive towards plastics, for example, strongly oxidizing liquids such as hot, fuming nitric acid. Other less powerful reagents promote premature brittle failure in plastics components under stress, a phenomenon known as *environmental stress cracking*.

The ageing of plastics can be the result of physical change or chemical reaction; the former includes the reorganization of crystal structure, free volume changes and material redistribution, such as the migration of plasticizer. Chemical ageing is very dependent on the environment, especially on temperature, oxygen, ozone, moisture, light (particularly ultraviolet radiation), and sometimes on biological factors. More than one of these agencies may be operative, leading to enhanced activity: for example, oxidation reactions are accelerated at high temperatures. Although some ageing processes progress inexorably, and can only be minimized by the elimination of factors which would worsen their effects, the incidence of other ageing phenomena can be reduced dramatically by certain additives. Antioxidants, to reduce oxidation, and UV stabilizers to protect against the effects of this radiation, are examples of great commercial importance.

1.3 Additives

Reference has already been made to antioxidants and UV stabilizers; these exemplify an important branch of plastics technology, the use of additives to modify the behaviour. Some additives are employed only in small concentration (0.1–1%), and are effective at this level, whereas others may constitute the major part of a plastics composition, being used, for example, to enhance the rigidity, or to reduce the cost. Additives used in low concentration include antioxidants, UV stabilizers, lubricants (to facilitate processing), slip and anti-blocking additives for film-surface modification, dyes and pigments. Additives used mainly to improve mechanical properties or economics include particulate fillers, fibres and gases, the last resulting in foams. Dispersion of particles of a solid with modulus higher than that of the plastics matrix inevitably results in a composite of higher rigidity. The increase in rigidity is smallest for fillers of spherical shape, and greatest for those of high aspect ratio, particularly long fibres. For all fillers the effects are proportional to the volume fraction of the additive and, as well as mechanical properties, improvements in dimensional stability and reduced thermal expansion and contraction are observed. Conversely, some additives are used to reduce the stiffness, notably plasticizers and rubbers.

Recently there has been increased use of surface coating materials, i.e. printing inks, metallization, or painting, to enhance the appearance and surface-related properties of plastics (e.g. barrier properties to gases or UV radiation, or increased abrasion resistance). Usually such treatment is accompanied by pretreatment to modify the surface of the polymer and to ensure adhesion of the surface coating to the substrate.

Composites offer the designer a means of extending the usefulness of plastics, but the behaviour can be very complex, requiring much more detailed definition of the properties. Thus, whereas the deformation of an isotropic material can be defined by two of the three moduli, anisotropic materials require a minimum of five engineering moduli for adequate description. A further disadvantage accruing from the use of composites is more difficult and less predictable processing behaviour; in particular, the shear viscosity is increased with filler content, even more than the modulus, and again, long fibres are the most effective. Composites involving fibre reinforcement are almost inevitably anisotropic in their properties. Commercially important composites are included in the later chapters, where appropriate.

1.4 Processing of plastics

As noted previously, an important factor favouring the use of plastics in any application is the easy shaping: for even the most recalcitrant polymer, shaping can be accomplished at moderate temperatures with equipment fabricated from conventional materials. There is the added advantage that many of the shaping processes are suited to automatic operation.

1.4.1 *Thermoplastics processing*

Thermoplastics comprise the majority of the market for plastics materials; their shaping will be discussed first. For the analysis of such processes, it is helpful to consider a number of 'unit operations', which provide a framework, or skeleton, on which the detail may be developed. The important stages are:

> Melting
> Mixing and homogenization
> Melt transport
> Primary shaping
> Secondary shaping
> Shape stabilization
> Finishing operations.

These operations are usually carried out sequentially and some, such as secondary shaping, may be omitted. Additional mixing may also be included, e.g. powder blending, where mixing is partially achieved prior to melting; however, final homogenization can only be attained by mixing in the molten state.

Melting, mixing and homogenization are frequently accomplished by an Archimedean screw rotating in a heated barrel, which is often an integral part of the processing equipment. Alternatively, a plastics compound can be made in a separate process, e.g. by batch operation in an internal mixer. Such a pre-mixed compound reduces the problem of achieving good dispersion of the components, but before shaping, the melt must also be at a uniform temperature and have experienced similar shear history, so that it will respond predictably and reproducibly in the subsequent shaping operations.

There are many ways of shaping a polymer melt, for example, a mould or an extrusion die; calendering is a further possibility. In shaping, the rheological properties of the polymer melt are relevant and determine the pressures necessary to move the melt. The *shear viscosity* of the melt is important, defining the shaping behaviour, and is frequently measured in quality-control tests, e.g. the melt flow index (MFI). The elastic properties are also important in shaping. Problems often arise due to interaction between the process and the structure of the polymer. Thus, the extrusion of complex profiles still defies accurate analysis, and most mouldings have residual strain, and possibly other defects, leading to anisotropic properties, which so far have not been predicted quantitatively. Furthermore, the shear response of the melt is modified by fillers and complicated by fibrous additives.

1.4.2 *Foams*

Foams are important because of their improved rigidity per unit weight, compared with solid plastics, and for the ease and versatility of their processing. In the design of foamed products, little advantage will be gained for section thicknesses less than 6 mm, i.e. by using existing injection moulds. Normal injection presses can be adapted for moulding foams, but they have unnecessarily high locking forces, together with restricted platen areas and plasticizing capacities. Polymer is injected into the cavity at low pressure and allowed to foam; the resulting pressure is low, and the locking force required is small. The platen size can be large, and large mouldings can be produced with small locking forces. The plasticizing unit should be of high capacity, to provide sufficient polymer melt for the shot. The special foam machines are considerably cheaper and more economical to run than conventional machines. The production of polyurethane foam materials is dealt with separately.

1.4.3 *Shaping of cross-linking plastics*

The basic stages, or unit operations, for forming cross-linking plastics are similar to those already considered for thermoplastics, but there are differences in detail and, more important, these differences allow some of the difficulties inherent in thermoplastics shaping to be circumvented. The fundamental difference between the processing of thermoplastics and cross-linking plastics is that, whereas chemistry is scrupulously avoided for the former, it is required for the latter. Thus thermoplastics melts have high viscosities, resulting from the high molecular weights necessary to achieve satisfactory mechanical properties. Robust equipment and high forces for manipulating the melt are necessary. By contrast, prepolymers for cross-linking systems can be of relatively low molecular weight and therefore of low melt viscosity; processing equipment can be lighter in construction, and forces very much lower.

Referring again to the scheme of unit operations detailed earlier in this section, the major difference between thermoplastics and cross-linking plastics is the shape stabilization stage; for thermoplastics, this is achieved by cooling, whereas for the latter, chemical reaction is involved, implying that the system is frequently heated at this stage. The shape-stabilizing reaction is essentially further polymerization carried out in the mould: heat is evolved and there is frequently considerable shrinkage. High levels of filler are added to reduce shrinkage and to cheapen the product; fibrous fillers and textiles are

particularly important and give composites of considerably enhanced properties. Glass fibres provide the usual fibrous reinforcement, although
asbestos still finds some application, and the newer carbon and Aramid™
fibres are used increasingly, often in mixed fibre systems. Cellulose fibres
contribute useful properties to composites, especially improving the impact
performance of PF mouldings.

1.5 Practical methods of processing

1.5.1 *Thermoplastics processing*

The techniques for primary shaping will be discussed first, to be followed by
consideration of the more important techniques of secondary shaping.

Extrusion is a continuous process which has the most economical output of
all plastics shaping methods. Extruded products are simple in shape and have
features in only two directions, being continuous in the third (extrusion),
direction. Common products manufactured by this process are pipe and other
rainwater goods, curtain rails, flat and tubular film, sheet, wire-covering and
fibres. The heart of this method is the screw, which consists of three sections,
fulfilling three functions: the feed and compression zone, the plasticizing zone,
and finally the metering section (Figure 1.3). Extruders may contain one, two,
or occasionally three separate screws in parallel.

Powder or granular material is fed from the hopper, through the throat and
on to the screw. Rotation of the screw transports the plastic along the cylinder
through the first zone, where it is compacted and trapped air is expelled.
Progressing down the cylinder, it is heated further by shear between the screw

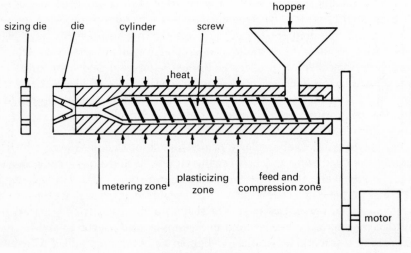

Figure 1.3 Extruder

and the cylinder walls, and by conduction of heat from the wall, so that the material is melted by the end of this zone. The final section of the extruder ensures the consistent and required output of molten polymer. The melt usually passes through a die, which defines a profile; this is cooled in air, or more frequently, in a cooling bath. A 'sizing die' is often included in the cooling train to rectify changes in profile of the extrudate after leaving the die.

Foams can be manufactured on standard extruders; the main problem being to control the amount of foaming on leaving the extruder. This is usually achieved with several sizing and cooling rollers. PVC foam for wood replacement is becoming an important market.

(i) *Injection moulding.*   Plastics material in granular or powder form is fed into the injection cylinder from the hopper, and is transported forward by rotation of the screw (Figure 1.4). Early injection presses used piston plungers to transport the melt, but from 1967 onwards, injection units have had reciprocating screw plungers. As the material moves down the cylinder, it is heated by shear between the cylinder wall and the screw and by conduction from the wall. (Band heaters are fitted to the outside of the cylinder). By regulating the power to the heaters, and the amount of shear on the plastic, it is uniform in temperature by the time it reaches the front end of the cylinder, and forms the melt pool. As more material is transported forward to the melt pool, the screw is forced backwards until a preset quantity of melt is available in the melt pool; this is commonly known as the shot weight, or more correctly, the shot volume. In spite of this, the shot weight is usually quoted, most frequently in terms of polystyrene. It must be remembered that this value must be corrected when the machine is moulding plastics other than PS, for example,

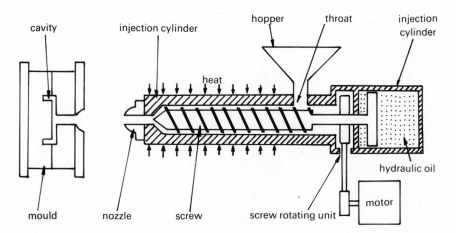

Figure 1.4 Injection moulder

the shot weight for PP is

$$\text{shot weight (PP)} = \text{shot weight (PS)} \cdot \frac{\text{density of PP}}{\text{density of PS}}$$

Thus the shot weight (PP) is some 15% smaller than that for PS.

The melt is then injected through the nozzle, sprue and runners into the relatively cool mould cavity by the forward movement of the screw, acting as a piston. Pressure is maintained on the melt while it cools in the cavity, until the entrance to the mould (gate) has solidified. The pressure is released and the screw starts rotating to prepare another shot, but the mould stays closed for a further time to solidify the material in the cavity before the mould opens, and the moulding is ejected. The mould then closes and the cycle is repeated.

Since most injection moulding machines can be run automatically, the process is suited to the making of large quantities of similar shapings at low cost. Complex mouldings may need retractable mould parts.

(ii) *Calendering.* This process was developed in the rubber industry, but more recently, the plastics industry has adopted the method to produce sheet or film, particularly from plasticized PVC. This is a process in which sheets or films of uniform thickness are produced by squeezing the molten polymer through a succession of rotating rolls. Modern calenders usually consist of four rolls or cylinders, which are temperature-controlled, in an inverted 'L' or inverted 'Z' configuration (see Figure 1.5). Molten polymer is fed to the first pair of rolls from an internal mixer, two-roll mill or extruder, with the roll separation controlling the feed rate. Subsequent roll separations (nips) refine the final product thickness.

An outstanding feature of this process is the very high output for a comparatively low energy input, compared with extrusion; further, it is especially suitable for heat-sensitive materials. Calendering is used to produce film and sheet, and for laminating plastics film to other substrates, including other plastics films, textiles, paper and metal foil.

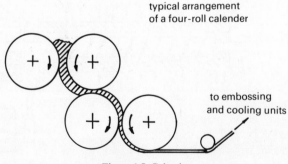

typical arrangement
of a four-roll calender

to embossing
and cooling units

Figure 1.5 Calender

(iii) *Rotational moulding.* This process is best suited to hollow components of simple shape, especially if they are required in small numbers, up to 1 000 per year. The process has been used for the manufacture of dustbins, 3 m³ storage tanks, small huts for road-workers, and boat hulls; it is also used to fuse plasticized PVC pastes, e.g. in the production of automotive parts, to include crash pads and armrest skins. The technique involves, as the title suggests, rotating a mould. The component is formed inside a hollow mould, which is heated, whilst being rotated on two axes, mutually at right angles. Four stages are involved in the moulding cycle: loading powder into the mould, melting and shaping, cooling and extracting the finished moulding. When the powder has been loaded into the cavity, the mould is closed, and it is heated and rotated. Rotation speeds are low, rarely exceeding 40 rpm on the secondary axis and 12 rpm on the primary axis. The mould is heated by air or oil, or by directing gas jets on to its exterior. The powder melts, forming a homogeneous layer of molten plastic on the inside of the mould, and becomes evenly distributed over the surface through gravitational forces. The heating and rotation are continued until all the powder has melted; the mould assembly is then transferred to a cooling system, involving either water or forced air. After solidification of the plastic, the mould is opened and the moulding removed.

Important methods of secondary shaping based on melt extrusion are outlined below.

(iv) *Tubular film.* Most film is manufactured by this process, in which a tube extruded vertically is simultaneously blown radially by introducing air into the tube (under pressure), and extended longitudinally by nip rolls running at a higher linear speed than the extrusion rate (see Figure 1.6). The nip rolls also serve to prevent the immediate escape of the air pressurizing the radial blow. The flow of air into the tube is controlled to give a constant bubble size and thus a consistent film thickness. The film is flattened by the rolls and passes to a wind-up unit.

(v) *Extrusion blow moulding.* Most small and medium-sized enclosed containers, such as bottles and small drums, are produced by this process; more recently, complex shapes, including car fuel tanks and coolant reservoir tanks, have been blow moulded.

A predetermined length of molten tubular extrudate, known as a parison, is introduced into an open mould (Figure 1.7). On closing the mould, the parison is trapped and welded at the top and bottom; air is then injected into the tube which is expanded and takes the shape of the cavity within the mould. Air pressure is maintained, keeping the moulding in contact with the cooled wall until it has solidified sufficiently to be form-stable. After the necessary cooling period the mould is opened and the component removed.

Blow moulding is strictly only suitable for manufacturing closed, hollow

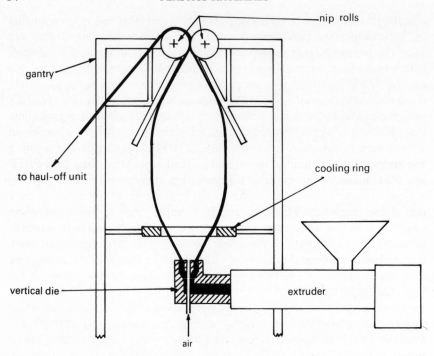

Figure 1.6 Blown film—tubular process

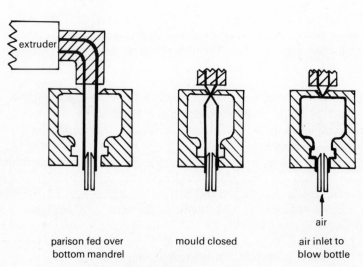

Figure 1.7 Blow moulding

objects, such as containers, although many complex shapes can be produced by ingenious adaptations of the technique. Machines and moulds are expensive, although less so than injection moulds, but fast production of small components to 1.5 m³ capacity containers is possible.

(vi) *Injection blow moulding.* In an analogous process, injection moulded preforms are reheated, and the secondary shaping achieved by blow moulding; this process is also the basis for the manufacture of *stretch-blown* bottles, in which balanced orientation is imparted to the walls of the bottle by extending the preform longitudinally, then blowing it radially. Bottles based on PETP and PVC made by this method dominate the market.

(vii) *Thermoforming.* This is a secondary shaping process which uses sheet or foil as feedstock, and forms a component by heating, followed by shaping, often using vacuum or compressed air. The sheet is first heated in an air oven, or more usually, by infrared heaters; single-sided heating suffices for sheet up to 3 mm thickness, otherwise double-sided heating is employed. When the sheet has softened, the component is shaped using a male or female mould, with a plug frequently involved to assist drawing (Figure 1.8). In the female mould, vacuum is applied to draw the material into the cavity (Figure 1.8a). If the draw is deep, much thinning will occur at the corners; however, this may be remedied if a plug is used to stretch the material before the vacuum is applied (Figure 1.8b). Alternatively, a male mould can be used on deep drawn sections, especially if compressed air is used to expand the sheet initially (Figure 1.8c). The mould can then be brought into contact with the stretched sheet, and the sheet collapsed to take the mould shape.

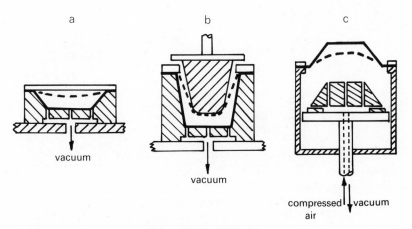

Figure 1.8 Thermoforming

1.5.2 *Processing methods for cross-linking plastics*

These methods involve the reactive processing of a prepolymer (or oligomer), or occasionally monomers, with a catalyst or a curing agent, as part of the shaping operation.

(i) *High-pressure moulding.* Important methods include the following techniques.

Injection moulding of thermosetting plastics, sometimes known as direct screw transfer, is very similar to the injection moulding of thermoplastics except that the plasticization is achieved at low temperature (60–90°C), and the curing occurs in the mould, which is at a temperature that produces rapid cross-linking (140–200°C). The material remains in the mould until it is cured sufficiently to be stable in shape (i.e. has *green strength*), when it can be demoulded although still hot. Injection moulding is used increasingly, particularly in respect of the high degree of automation possible in the process. Typically phenol-formaldehyde novolaks and amino-plasts are processed this way. A variant is the shaping of dough moulding compounds (DMC), based on catalysed and glass-filled polyester materials.

Compression moulding is carried out on moulding powders or sheet moulding compounds (SMC). The presses are vertical, with the hydraulic cylinder above or below the horizontal platens. The mould is heated to between 160 to 200°C, being thermally insulated from the platens (Figure 1.9). The required amount of prepolymer (mixed with catalyst or curing agent) is placed into the open mould; it is then closed and pressure is applied to cause the material to flow around the cavity. The mould remains closed until the

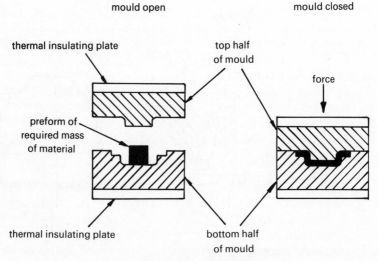

Figure 1.9 Compression moulding

prepolymer has cross-linked. The non-homogeneity of the polymer melt during shaping can lead to problems in the finished product, e.g. poor fusion or dislodged inserts. For sheet materials, the charge is cut and placed in the mould cavity and pressure is applied to the hot mould to force the material to take up the shape of the cavity. Pressure is maintained until curing is complete.

Transfer moulding is a variant on compression moulding, in which there is a reservoir of catalysed molten resin, an aliquot of which is 'transferred' to the mould cavity at an appropriate point in the cycle, in a fairly homogeneous state.

(ii) *Low-pressure moulding.* Methods have been developed particularly for glass-reinforced polyester (GRP) systems. However, they are generally applicable to most reactive dual component systems, in which liquid prepolymers can be easily mixed and dispensed into moulds, wetting out fibre reinforcement, processing being carried out at room temperature. In the simplest case an open mould is employed, made from timber, aluminium or GRP itself (polyester or epoxide-based), and having a smooth surface finish. To this is applied, in sequence, mould release agent; gel coat (free of glass fibre); glass tissue impregnated with resin (optional); resin with major reinforcement, repeated to build up the required thickness (hence eventual required product strength); tissue impregnated with resin (optional); and the final gel coat. Catalyst (the reaction initiator) is mixed with the resin prior to application so that it cross-links under ambient temperature conditions, although the reactivity can be readily adjusted to allow adequate working time before gelling. Two major processes have been developed: hand lay-up and spray methods.

In the hand lay-up process, the reinforcing 'mat' is cut to shape and disposed in the mould manually prior to impregnation with catalysed resin, applied with brush and squeeze-roller. In the spray process, chopped glass fibres are incorporated in a spray of catalysed resin applied to the mould. Both techniques allow the manufacture of large continuous mouldings limited only by the size of the mould.

A variant of low-pressure moulding is catalysed resin injection into matched moulds in which the reinforcement has been preplaced prior to mould closure; vacuum applied before resin injection assists mould filling. All these techniques depend on the easy flow of the prepolymer and good wetting of the reinforcement, but preclude the addition of much cheap particulate filler.

(iii) *Dual component dispensing techniques.* Many thermosetting plastics, (e.g. polyurethanes, epoxides, silicones, modified polyesters, phenol-formaldehyde and amino-plasts), and some thermoplastics (e.g. modified PA 6 and certain acrylics), can be shaped directly from two (or more) low-viscosity liquid monomeric components, which have to be mixed in precise stoich-iometric proportions immediately before processing. Several dual component

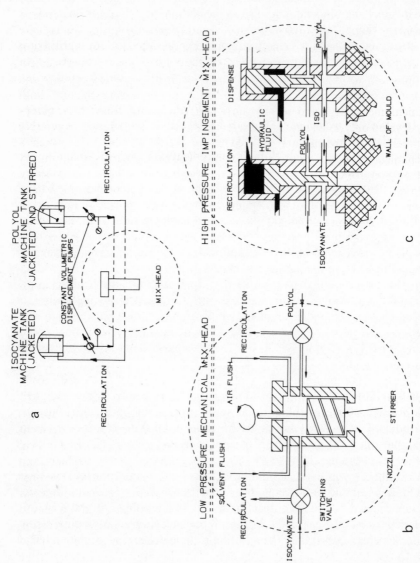

Figure 1.10 (*a*) Schematic diagram of dual component dispensing equipment. (*b*) Low-pressure mechanical mix-head. (*c*) High-pressure impingement (or contra-flow) self-cleaning mix-head

dispense methods with accurate metering of ingredient outputs, and subsequent ratioing and mixing of ingredient streams, have been developed (see Figure 1.10).

Polyurethanes are the best example of materials using *low-pressure dispensing*. Here a polyol blended with catalyst, blowing agents, etc., is held conditioned in a tank, and the isocyanate in a second tank. An ingredient is pulled from its tank and metered using a constant volumetric displacement pump, which when combined with restrictions in-line develops pressure and velocity in that ingredient stream. By adjusting the output on each pump, both a stoichiometric ratio and output rate (a *shot*) can be established. Gear pumps will develop only a few bars pressure, enough to feed a mix-head. To produce polymer, both streams are allowed to enter the mix-head, in which a mechanical propeller rotating at 5000–15000 rpm shears and intimately mixes the ingredients, which will then rapidly exit through the head's nozzle (Figure 1.10a). The low velocity with which the ingredients leave the head allows *pouring* without splashing. Once a shot is fired, three-way valves shut ingredients off from the mix-head, and residual ingredients must be flushed out using a suitable solvent followed by an air blast; this leads to waste problems. The low-pressure equipment is used for open- and closed-pour moulding and the manufacture of foam slab-stock.

High-pressure dispensing is achieved using any of a variety of piston pumps (e.g. radial type), where pressures of 150–250 bars are developed (lower for polyamide reactants). A typical high-pressure mix-head consists of cylinder with inlet and return ports, and a hydraulically operated ram (Figure 1.10b). On recirculation, each ingredient stream from the pump passes along a small groove in the side of the ram, and back to its tank. When a shot is fired, the ram retracts, clearing the inlet ports but sealing the return ports. High pressure, combined with small cross–sectional area at the inlet port, means that each stream will enter the mixing chamber with high velocity and kinetic energy. The energy will be mostly converted into complete and intimate impingement mixing of the two streams: each stream needs to reach a minimum critical Reynolds number for good mixing—100 or more for cellular products, and 150 or more for non-cellular products. The residence time in the mix chamber is very short and the mixed ingredients will rapidly leave the head's nozzle at $1-2\,\mathrm{m\,s^{-1}}$. Normally, ingredients can only be safely *injected* into a mould cavity. The high-pressure equipment is used predominantly in reaction injection moulding (RIM) processes, although modification of the mix-head permits spray processing, and open- and closed-pouring. The viscosity of the ingredients should not be above 5 Pa s for successful pumping and mixing, although heating ingredients can be an effective way of reducing viscosity. Fibre and particulate filled systems can only be processed on modified high-pressure equipment. Since low-viscosity mixed ingredients are injected into the mould cavity, filling is achieved with relative ease, the maximum pressure in the mould being 3 bars for some foamed systems, and less than 1 bar for most other systems (compare to approximately 1 500 bars for injection moulding thermoplastics). As a result, cheaper weaker tooling materials (e.g. aluminium and kirksite alloys, reinforced epoxide resin), along with smaller clamping presses can be employed; toggle or G-clamps will even suffice with smaller moulds. Equally, maximum moulding weight possible with RIM equipment can exceed 100 kg, with clamping forces required to hold the mould closed being less than a tenth of those needed for thermoplastic injection moulding. However, steel tools have to be used in long production runs.

1.6 Interactions between shaping process and plastics materials

There are two events in the shaping process which would be expected to influence the properties of the product via their effect on the materials being processed. These are movement of material to accomplish shaping, and change of phase to achieve stabilization of shape. There may be additional interactions, such as if homogenization is not achieved prior to shaping, or when additional material is packed into a mould to compensate, in part, for shrinkage. However, most problems in practice arise either from failing to conform to the principles already discussed in this chapter, or from failure to appreciate the significance of the two factors given above.

A consequence of the movement of polymer during shaping is a tendency of the polymer chains to orient in the direction of flow. Further, if anisotropic fillers are incorporated in the formulation, these will also be influenced by the flow geometry. These phenomena are inevitable and can only be reduced to acceptable levels by attention to the detail of the shaping process, or by harnessing orientation to reinforce the component in the direction of maximum stress by appropriate mould or part design. Orientation, and consequently anisotropy of mechanical properties, is frozen into the component by the shape stabilization stage. Orientation generally leads to *birefringence*, the occurrence of different refractive indices in the material for different directions of viewing. At the factory-floor level, this can be a useful effect since, in its simplest form, it can be seen by placing a transparent moulding between crossed Polaroid TM sheets. For translucent and opaque objects the procedures are a little more complicated.

The stabilization stage, necessarily involving a phase change from liquid to solid, has several consequences.

(i) *Shrinkage.* Thermoplastics have higher specific volumes in the melt than in the solid state. This is true for amorphous thermoplastics and even more so for crystallizing materials, but some compensation may be obtained by packing the mould under high pressure, thus taking advantage of the limited compressibility of the plastics melt. It is sometimes possible to provide more plastics melt to compensate for the shrinkage, as for example in specially designed cooling dies for extrudates of thick section or by hot runners in injection moulding. These palliatives are not effective for cross-linking (thermosetting) plastics, where the major shrinkage results from the poly-merization reaction. Shrinkage can be reduced by inert fillers and by shrink control additives: these are additives, frequently polymeric, which are soluble in the uncured resin but insoluble in the cured, cross-linked plastic. The precipitation of the additive results in an increase in volume, compared to that of the solution. The contribution of shrinkage to tolerances is considered in greater detail in Chapter 2.

(ii) *Crystallinity.* The crystallinity level attained on cooling a plastics melt depends on many process and materials factors. A very important factor is the rate of cooling, rapid chilling leading to lower levels of crystallinity with consequent decrease in mechanical strength, modulus and (usually) an improvement in toughness.

(iii) *Crystalline texture.* Cooling of a plastics melt can only be accomplished by heat transfer by conduction to a heat sink. On simple heat-transfer grounds alone, therefore, cooling times can be minimized by using moulds or cooling baths, as cold as practicable. This frequently leads to a quenched zone of reduced crystallinity and finer crystalline texture which hitherto has been assumed to be beneficial to mechanical properties. However, there is mounting evidence that some mechanical weakness resides in the boundary between this region of fine texture and the bulk of coarser crystalline material. The use of cold moulds in injection moulding of thermoplastics may restrict flow unduly, leading to short mouldings or to unacceptable levels of residual strain.

Further reading

Polymer science

Bassett, D.C., *Principles of Polymer Morphology*, Cambridge University Press (1981).
Billmeyer, F.W., *Textbook of Polymer Science* (3rd edn.), Wiley-Interscience, New York (1984).
Cowie, J.M.G., *Polymers: Chemistry and Physics of Modern Materials*, International Textbook Co. [Blackie Publishing Group], Glasgow and London (1973).
Haslam J., Willis, H.A. and Squirrell, D.C.M., *Identification and Analysis of Plastics* (2nd edn.), Iliffe [Butterworth], London (1972).
Parker, D.B.V., *Polymer Chemistry*, Elsevier-Applied Science, Amsterdam (1974).
Rodriques, F., *Principles of Polymer Systems* (2nd edn.), McGraw-Hill, New York (1982).

Plastics properties

Brown, R.P., *Handbook of Plastics Test Methods*, Godwin [Longman Group], London (1981).
Brydson, J.A., *Plastics Materials*, Iliffe [Butterworth], London (1982).
Blythe, A.R., *Electrical Properties of Polymers*, Cambridge University Press (1979).
BS 4618, Recommendations for the Presentation of Design Plastics Design Data. British Standards Institution, London.
Crawford, R.J., *Plastics Engineering*, Pergamon Press, Oxford (1987).
Hall, C., *Polymer Materials: An Introduction for Technologists and Scientists*, Macmillan, London (1981).

Hearle, J.W.S., *Polymers and their Properties:* Vol. 1: *Fundamentals of Structure and Mechanics*, Ellis Horwood, Chichester (1982).

Ogorkiewicz, R.M., *Thermoplastics: Properties and Design*, Wiley-Interscience, New York (1974).

Ogorkiewicz, R.M., *The Engineering Properties of Polymers* (Engineering Design Guides No. 17), Oxford University Press (1977).

Turner, S., *Mechanical Testing of Plastics* (2nd edn.), Iliffe [Butterworth], London (1983).

Polymer manufacturing and processing

Becker, W.E., *Reaction Injection Moulding*, Van Nostrand Reinhold, New York (1979).

Brown, R.L.E., *Design and Manufacture of Plastic Parts*, Wiley-Interscience, New York (1980).

Brydson, J.R., *Flow Properties of Polymer Melts* (2nd edn.), Godwin [Longman Group] (1981).

Cogswell, F.N., *Polymer Melt Rheology*, Godwin [Longman Group] (1981).

Fenner, R.T., *Principles of Polymer Processing*, Macmillan, London (1979).

Radian Corporation, *Polymer Manufacturing: Technology and Health Effect*, Noyes Data Corporation, Park Ridge (1986).

Tadmor, Z. and Gogos, C.G., *Principles of Polymer Processing*, Wiley-Interscience, New York (1979).

2 Fundamentals of design

Design is an exercise embracing many considerations: mechanical performance and cost are obviously relevant, and the latter has many constituent elements, of which mass cost (or more appropriately, volume cost) is one of the most important. Data are given in Table 2.1; however, manufacturing cost can modify the overall cost very considerably, and as the cost of plastics is very unstable at present, data in Table 2.1 must be regarded as only approximate. Different shaping methods are employed for plastics and each has an associated processing cost depending mainly on the production quantity required. For instance, the initial capital cost of equipment for injection moulding and extrusion can be very high, whereas the cost for equipment in GRP hand lay-up may be merely the cost of a wooden former. Expensive processes can only be justified if the production quantity is large, allowing the costs to be absorbed, thus making the processing cost per item acceptable.

2.1 Engineering design

When a component is subjected to an external load, stresses are created within the component and it will deform or deflect. In metals and other materials, the stresses and deflections can be calculated because the material behaves linearly (in the elastic region) when an external load is applied (Figure 2.1). This, commonly known as Hookean or linear elasticity, can be used for many design applications for metals because the deformation is proportional to the load for a constant cross-section area. When most plastics are subjected to a similar situation, the relationship is non-linear (Figure 2.2). This response to applied force is generally described as viscoelastic because the material behaves partly as a fluid of very high viscosity and partly as an elastic solid. The traditional engineering relationships used by the designer based on Hookean principles are inadequate, but if the plastic is assumed to behave linearly, the designer can calculate the stresses and deflections for short-term loads.

If the load is maintained for a long period the material will continue to flow as a high-viscosity fluid (Figure 2.2). This introduces the important concept of 'creep' into plastics design. The amount of creep will depend on the temperature of the component under load, the length of time the load is applied, and the stress level (Figures 2.3, 2.4). This same information can be presented in various other ways; either by a stress vs. strain graph with curves

Table 2.1 Usage and mass cost of plastics materials

Plastics materials	Usage (Tonnes × 1000) 1984	1985	1986	Price Band[†]
Thermoplastics				
Low density polyethylene (LDPE)/	510	580	610	A
Linear low density				
polyethylene (LLDPE)				A/B (C_8 copolymer)
High density polyethylene (HDPE)	210	250	275	A
Polypropylene (PP)	297	330	370	A (homopolymer)
				B (copolymer)
Orientated polypropylene				
(OPP) film	39	40	43	G
Poly(vinyl chloride)(PVC)	444	450	484	A
				B (emulsion)
Polystyrene (PS)/styrenics	146	148	156	B (crystal)
				B (HIPS[1])
Expanded polystyrene (XPS)	31	29.5	33	C
Acrylonitrile-butadiene-styrene (ABS)	53	55	58	E
Polyamide (PA)	20	22	23	G (PA 6 and 6.6)
				I (PA 11 and 12)
Poly(oxymethylene) (POM)	10	10.8	12	D/E
Thermoplastics polyesters	37	41	45	D/E (PETP[2])
				G (PBTP[3])
PETP film	23	23.7	24.4	G
Poly(tetrafluoroethylene) (PTFE)	1.25	1.35	1.4	I
Poly(methyl methacrylate) (PMMA)	27	28	29	F
Poly(phenylene oxide)/				
Polystyrene (PPO/PS)	6.0	5.8	—	G
Thermosets				
Phenol-formaldehyde (PF) resins	49	50	49	D-F (Novolak resins)
				B (Resol solution)
				D/E (Resol–100% solids)
PF (moulding grades)	14	14	12	B
Amino-plasts	122	125	131	C/D (UF[4] moulding powders)
				C/D (MF[5] moulding powders)
				A-D (MF impregnation
				resins and solution)
Unsaturated polyester resins (UPR)	50	51	53	E/F
Epoxides*	17	17.5	18.5	F
Polyurethanes* (PU)	91	94	100	D/E (foams)
				E-G (elastomers/resins)
				G (thermoplastics grades)

Notes

[†]*Price band* is based on the average cost of tonne lots for the UK in 1987.

A £400–£600	F £1500–£2000
B £600–£800	G £2000–3000
C £800–£1000	H £3000–£5000
D £1000–1250	I > £5000
E £1250–£1500	

Abbreviations

    (1) HIPS High impact polystyrene,
    (2) PETP Poly(ethylene terephthalate),
    (3) PBTP Poly(butylene terephthalate),
    (4) UF Urea-formaldehyde,
    (5) MF Melamine-formaldehyde.

*Epoxide and polyurethane materials are available in many chemical forms. Prices indicated based on the *general-purpose*, rather than *specialist* grades.

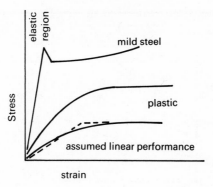

Figure 2.1 Typical stress vs. strain curves for metal and plastic

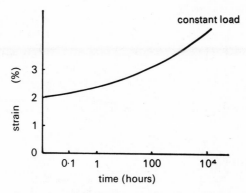

Figure 2.2 Schematic representation of a creep curve

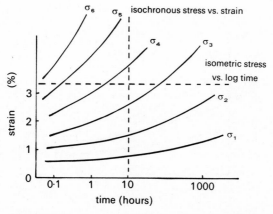

Figure 2.3 Creep curves at 20°C for various stress levels

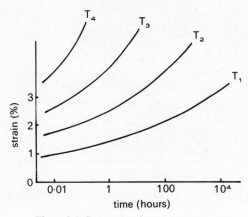

Figure 2.4 Creep at various temperatures

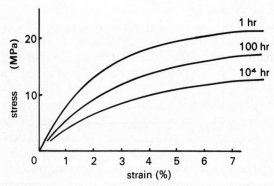

Figure 2.5 Isochronous stress vs. strain curves derived from creep data

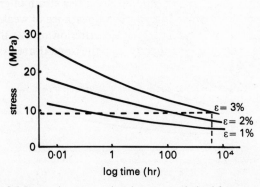

Figure 2.6 Isometric stress vs. log time curves derived from creep data

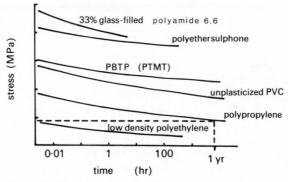

Figure 2.7 Long-term strength data for some thermoplastics

at different times (Figure 2.5: this is an isochronous stress vs. strain curve) or by a stress vs. log time curve to show the maximum design stress (Figure 2.6: an isometric stress vs. log time curve). Both curves can be derived from the creep data at 20°C, as indicated in Figure 2.3.

The maximum strength will vary with time (Figure 2.7) and temperature. A designer, faced with the problem of designing a component, can use a criterion either of maximum stress or of maximum deflection. Strain-limited design is adopted for ductile plastics, whereas strength is used as the basis for design with brittle materials and for brittle failure in otherwise ductile materials. For example, if a tie bar made from propylene homopolymer were to be continuously loaded at 20°C for one year, what would be the design stress for the tie bar? The designer could use a design criterion of maximum long-term strength of 18 MPa (Figure 2.7) and apply a safety factor of 1.5, to arrive at a design stress of 12 MPa. The value of the safety factor will depend on a number of variables (such as shape, position of load, stress concentrations, weld or joins), so for a simple bar a safety factor of 1.5 can be assumed to be adequate. However, from creep data, after one year at a stress of 12 MPa, the bar will have extended more than 4% of its original length. This will lead to stress-whitening, which is unattractive, but additionally a strain of greater than 4% may be unacceptable as well as giving the appearance of weakness. The part could be designed with a maximum strain criterion of 3%; then the design stress of 9 MPa can be obtained directly from the isometric stress vs. time curve

Table 2.2 Upper strain limits (%) for representative plastics

Glass Reinforced Polyamide 6.6	1.0
Acetal copolymer	2.0
Polypropylene	3.0
Polypropylene with welded joints	1.0
ABS	1.5
Polyethersulphone	2.5

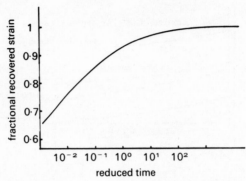

Figure 2.8 Fractional recovered strain vs. reduced time (PBT at 20°C)

(Figure 2.6). Although this value is lower than the previous design stress derived from maximum strength, it is likely to produce a more acceptable product.

The maximum strain criterion is becoming popular for plastics materials subjected to long-term loads; upper strain limits for representative plastics are suggested in Table 2.2. Thus the appropriate design stress at a particular time and temperature can be found in the relevant creep data.

When a load is removed from a metal part, the recovery from strain, if it occurs, is instantaneous. With plastics, much of the strain is recovered reasonably quickly but a proportion is recovered only after a period of time. Thus, recovery, like creep, is time-dependent. Recovery data are provided as fractional recovered strain vs. reduced time (Figure 2.8), where:

$$\text{Fractional recovered strain} = \frac{\text{Strain recovered}}{\text{Total creep strain when load is removed}}$$

and

$$\text{Reduced time} = \frac{\text{Recovery time}}{\text{Duration of creep period}}$$

The effect of intermittent loading on creep properties is less than for continuously-loaded applications, but intermittent loading may introduce fatigue; the resistance to fatigue varies between materials. Examples of intermittent loading are shown in Figure 2.9 for different stresses and loading and unloading periods.

Impact strength is often a very important design requirement and is affected by three principal factors–temperature, stress concentration and materials factors. It is important to remember that a tough material may behave in a brittle manner at low temperature; however, stress concentrations are the most common source of failures in practice because of the very high stresses induced at corners, holes, keyways, etc.. An example of the effect of stress

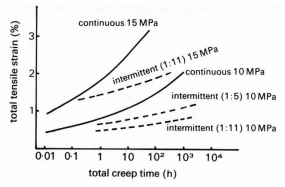

Figure 2.9 Creep under conditions of intermittent loading (propylene copolymer at 20°C)

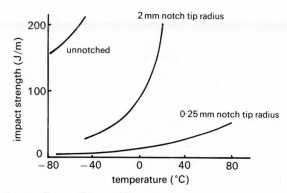

Figure 2.10 Impact strength vs. temperature (rigid PVC)

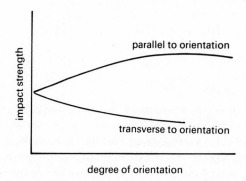

Figure 2.11 Effect of orientation on tensile strength of polystyrene injection mouldings

Table 2.3 The impact strength of thermoplastics (ICI Technical Service Note G123)

Material	Temperature (°C)							
	−20	−10	0	+10	+20	+30	+40	+50
Polystyrene	A	A	A	A	A	A	A	A
Acrylic	A	A	A	A	A	A	A	A
Glass-filled polyamide 6.6 (dry)	A	A	A	A	A	A	A	B
Poly(4-methylpentene)	A	A	A	A	A	A	A	AB
Polypropylene	A	A	A	A	B	B	B	B
Poly(methyl methacrylate) (craze-resistant)	A	A	A	A	B	B	B	B
Poly(ethylene terephthalate)	B	B	B	B	B	B	B	B
Polyacetal	B	B	B	B	B	B	C	C
Rigid PVC	B	B	C	C	C	C	D	D
CAB	B	B	B	C	C	C	C	C
Polyamide 6.6 (dry)	C	C	C	C	C	C	C	C
Polysulphone	C	C	C	C	C	C	C	C
High-density polyethylene	C	C	C	C	C	C	C	C
Poly(phenylene oxide)	C	C	C	C	C	CD	D	D
Propylene–ethylene copolymers	B	B	B	C	D	D	D	D
ABS	B	D	D	CD	CD	CD	CD	D
Polycarbonate	C	C	C	C	D	D	D	D
Polyamide 6.6 (wet)	C	C	C	D	D	D	D	D
PTFE	BC	D	D	D	D	D	D	D
Low-density polyethylene	D	D	D	D	D	D	D	D

Notes
  A. *Brittle*: specimens break even when unnotched
  B. *Notch brittle*: specimens brittle when bluntly notched but do not break when unnotched.
  C. *Notch brittle*: specimens brittle when sharply notched.
  D. *Tough*: specimens do not break even when sharply notched.

concentrations is shown in Figure 2.10. It is the designer's concern to radius corners and avoid stress concentrators in an area susceptible to impact or other loads. The effect of orientation is to introduce anisotropic properties to the component which will reduce the impact strength (Figure 2.11).

A list of plastics materials in order, from the most brittle to the most ductile is given in Table 2.3. However, it is important to appreciate that toughness in impact does not imply good fatigue behaviour.

2.1.1 *Fracture mechanics applied to plastics*

Sometimes plastics products, such as toys and domestic commodities, fail prematurely. This may be caused by poor materials choice, bad processing, poor-quality material, poor design, or for other reasons. In items such as pipe and storage vessels, failure can be catastrophic, and designers seek to produce safer products. Fracture mechanics aims to assist the designer in understanding the mechanics of crack growth by answering the following questions:

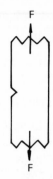

Figure 2.12 A crack subjected to opening forces

(i) What will be the residual strength of a cracked part with a change in crack size?
(ii) What is the critical crack size for catastrophic failure at the expected service load?
(iii) How long will it take to grow a minor crack to a critical crack size?

Most of the theory developed to date has been concerned with the behaviour of cracks or crack-like defects produced by opening forces (Figures 2.12, 2.13) applied to linear elastic brittle materials. More complex analyses have been developed for cracks induced by shearing or tearing, but these are beyond the scope of this text. On first consideration, the relevance of a fracture mechanics approach to plastics which are predominantly ductile may be questioned, but further reflection confirms that most unexpected failures are brittle in nature.

For tensile stress fields, the basis of the fracture mechanics calculation is the way the stress is concentrated in the region of the crack or defect, described by the stress intensity factor, K_I. Consider a thin plate of infinite size (Figure 2.13a), containing a through-thickness crack of length $2a$, subjected to a uniform uniaxial stress, σ, measured remote from the crack (the stress is normal to the crack growth). The stress is enhanced near the crack tip (stress

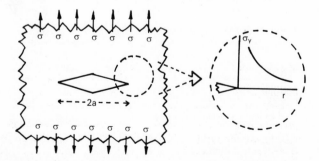

Figure 2.13 Further analysis of a crack subjected to opening forces

concentration), this local stress, σ_y (see Figure 2.13b), being given by

$$\sigma_Y = K_1[2\pi r]^{1/2}$$

where r is the distance from the crack tip along the direction of crack growth, and

$$K_1 = \sigma[\pi a]^{1/2} (= \text{linear elastic stress intensity factor})$$

These equations show that the local stress is proportional to the applied stress and varies with the square root of the crack size. K_1 is independent of the material from which the plate is made, provided that it is linear elastic. K_1 does depend on the size and location of the flaw (and on the applied stress). The location of the crack in relation to the plate geometry is incorporated in a modifying factor, Y.

$$K_1 = Y \cdot \sigma \cdot [\pi a]^{1/2}$$

Values of Y can be obtained from appropriate texts [1].

The considerations so far have not taken account of the properties of the material under stress, the only requirement being that it is linear elastic. The material will fail by very rapid crack growth when K_1 reaches a critical value for the material K_{IC}, often called the fracture toughness. Typical values for fracture toughness for different plastics subjected to short-term loading are given in Table 2.4.

The relationship between applied stress and critical crack length for catastrophic failure under an opening force is shown in Figure 2.14 for a range of values of fracture toughness: the importance of this parameter is clearly seen. Fracture toughness is affected by factors which affect strength, including time under load, temperature, nature of the environment, effect of processing conditions and molecular weight. Thus deciding whether a product is likely to fail depends on the relevant value of the fracture toughness, the applied stress, the flaw size and the dimensions of the article, together with the nature and location of the crack, as expressed by Y.

However, it is not possible for the applied stress to exceed the failure stress, σ_F, for the material. If σ_F is applied to the product which does not contain a

Table 2.4 Typical values of fracture toughness at 20°C in air.

Material	K_{IC} (MN/m$^{3/2}$)
Epoxide	0.6
Polystyrene	1.0
Poly(methyl methacrylate) (cast sheet)	1.6
Polycarbonate	2.2
PVC (pipe compound)	2.3
High-density polyethylene	3.0
Polyamide 6.6	3.6

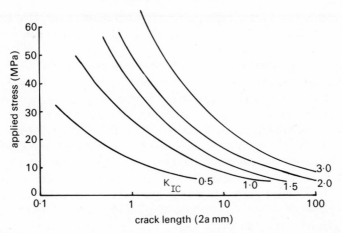

Figure 2.14 Relationship between applied stress and critical crack length for catastrophic failure for various values of fracture toughness

deliberately introduced flaw, then it is possible to calculate the inherent flaw size a_{IFS} that the material seems to have:

$$a_{IFS} = K_{IC}^2 \cdot [\pi \sigma_F^2 Y^2]^{-1}$$

It would seem that if there were a crack in a structure there would be no reason to be concerned, provided its size was smaller than the critical crack size. This is not true–any crack in a body under stress will gradually increase in size with increasing speed, until catastrophic failure occurs when $K_1 = K_{IC}$. The crack growth rate, a, is given by:

$$\dot{a} \simeq C K_1^n$$

for materials under constant tensile stress, with C and n constants, depending on the materials and test conditions. Experiments have shown that there are frequently three distinct areas of crack growth: slow crack growth (smooth fracture surface) followed by intermediate growth and finally fast growth (rough fracture surface). For each region, different values of C and n apply.

The time taken for a crack to grow can be found by integrating the crack growth equation between the original size of the crack and the new size of the crack.

2.2 Design limitations imposed by processing method

Before product design can commence in earnest, the designer must have considered which processing method to employ, since this has a direct bearing on the design.

GRP hand lay-up methods are suitable for simple open design shapes (chair shells, covers, etc.) which can be easily removed from the wooden or metal former; otherwise, more complex large enclosed shapes (such as truck cabs) can be manufactured if the former can be dismantled after moulding. Generous tapers of 2° or more are necessary to assist removal from the mould. Localized highly stressed areas can be made thicker and stronger by appropriate placement of the reinforcement.

Rotational moulded parts (dustbins, barrels, etc.) are usually enclosed sections. Generous radii are necessary to assist flow into corners; sharp corners would make the section thin, as well as acting as stress concentrators, making them very weak. Normally the bulk powder volume should be no more than 50% of the cavity volume, otherwise bridging from one side to another will occur. Ribs, lugs, surface features and textured surfaces are possible for rotationally moulded parts. Mouldings produced by rotational casting have very little residual orientation, and the long cooling time will tend to produce highly crystalline mouldings.

Sheet thermoformed parts (vacuum forming) are usually simple in shape and have all the features in two dimensions (baths, small boat hulls, body panels, dispenser cups, etc.). The main criterion for thermoformed parts is the length of draw to the original sheet thickness, because this will directly affect wall thickness. Thinning will occur at the corners, so generous radii are necessary to reduce the inherent weakness at these points. Corrugations can be used to increase the stiffness of the part perpendicular to the corrugations. Orientation is induced in the direction of draw, and remains in the finished moulding. Reproduction of mould image is limited, e.g. sharp corners becoming rounded, since thermoforming does not involve a plastics melt; reproduction becomes worse through the thickness of the moulding.

Blow moulding is used to manufacture enclosed shapes (bottles, barrels, containers, etc.) where the shape is simple. Screw threads can be formed on the top, corrugated sections can be used to stiffen the product, and handles can be incorporated. Thinning will occur on the corners of the base and also on the corners in the sides of square bottles, but can be reduced by blowing a profiled parison. Some orientation is created in the blowing the parison, and is largely retained in the finished product (e.g. at a maximum in the vertical walls of a bottle).

Extruded parts are normally one-dimensional and of a simple cross-sectional shape (curtain rail, rainwater gutters, pipe, or films which are produced from a very wide die and hauled off at a faster rate to produce orientation). In the extrusion of sections, the part thickness will affect distortion, and may lead to sinking and voiding. The polymer chains are extended in the direction of extrusion and the properties are correspondingly anisotropic.

Thermoplastic injection moulding is by far the most common moulding process in use, as well as being amongst the most complex. Normally, injection moulded parts have features which are formed parallel to mould opening (two-dimensional features). Features perpendicular to mould opening are possible but these are formed using complicated tooling arrangements. This manu-facturing method is suitable for complex parts provided they are not fully enclosed.

Reaction injection moulding involves an additional factor in process control, that of shaping of reacting monomers, which have to be dispensed in controlled stoichiometry to ensure complete polymerization and cross-linking. With the exception of fibre-filled materials, orientation is limited in RIM products.

2.3 Product design

Mention has been made previously of the two different classes of plastics materials—thermoplastics and thermosetting plastics—and to additives affect-ing mechanical properties (fillers, glass reinforcement, blowing agents). Each of these subgroups has different design principles and the additives alter the way in which the component should be designed.

2.3.1 Thermoplastics

Thermoplastic injection moulding of constant thin cross-sectional thickness is necessary to avoid surface sink marks and voids within the core (Figure 2.15a). This will also assist in reducing distortion and result in more accurate tolerances for a part. The wall thickness of ribs and bosses must be kept to less than 2/3 of the main wall thickness to prevent sink marks appearing on the surface (Figure 2.15b). If thick ribs are necessary, it may be possible to disguise the sink marks by surface features (Figure 2.15c, d). Flat plates are susceptible to warping; additional ribbing or corrugating the surface will produce a much more rigid structure (Figure 2.15e).

The melt in a mould cavity will be forced to flow around holes and slots, joining again after the hole to form a 'knit line' or 'weld line'. The gates should be positioned so as to reduce the number of weld lines to a minimum, because weld lines are potential weak points in any moulding. Similarly, in extrusion, the melt is sometimes forced to flow around restrictions in the die, and weld lines are formed, which again can create a weakness in the product. Thermoformed and blow moulded parts have flat sections or straight sides which are susceptible to warping and bending. It is again possible to reinforce these sections by using a corrugated shape.

2.3.2 Thermosetting plastics

These materials are shaped by reaction processing, such as screw transfer injection or compression moulding, although glass reinforced polymers (GRP) are frequently moulded using hand lay-up or spray techniques. The injection techniques and compression moulding of cross-linking plastics do not have

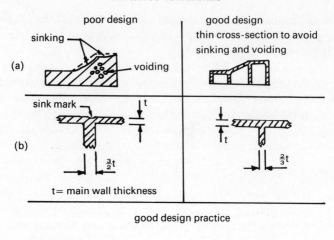

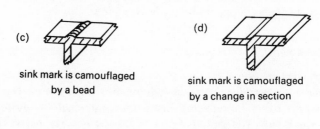

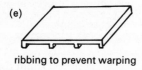

Figure 2.15 Common moulding faults and some corrective measures

the same limitations as thermoplastics injection moulding regarding wall thickness, rib size, etc. In part, this is due to the very low viscosity of melts or mixed monomers when flowing around the mould cavity, compared to that of thermoplastic melts. Mouldings with thick sections and with varying sections do not create any problems except that cure time may be affected by thickness. A generous draft angle (approximately 2°) is necessary to assist the ejection of the moulding from the mould.

2.3.3 Foams and reinforced plastics

Occasionally, the attractive properties of polymers fall short of the needs of a designer. One way of altering these is to include an additive to improve properties to those required. Blowing agents and reinforcing fillers will change the basic properties of a polymer: as well as the mechanical properties, the melt flow properties are changed, and this may affect flow length, mould filling and mould cycle times as well as the mechanical properties indirectly, for example by increasing anisotropy.

(i) *Foams*. The manufacture of foamed parts had been restricted to the cheaper polymers and polyurethanes, although the range now includes engineering thermoplastics. Foams are often classified by density: low-density foams ($10–100\,kg\,m^{-3}$) are used in packaging, cushioning materials and thermal insulation; medium-density foams ($100–600\,kg\,m^{-3}$) are used for footwear, furniture and impact restraint pads; high-density foams (greater than $600\ kg\,m^{-3}$), sometimes referred to as 'structural' or 'microcellular' foams, are used in many 'rigid' applications, for example washing-machine tubs, television panels, business-machine covers, vehicle panels and artificial leathers. In general, the advantages of a structural foamed part over an unfoamed plastics component are:

Cost reduction
Increased rigidity for the same polymer weight
Reduced weight
Possibility of moulding thick and thin sections.

Furthermore, the manufacture of foams is attractive because it is possible to reduce manufacturing costs. This is possible because the mould has greater cross-sectional thickness, which reduces the resistance to the flow of the melt in the mould and consequently requires much less injection pressure. Also resulting from the reduced injection pressure for thermoplastic foams, the mould clamping force is much less and needs only to keep the two halves of the mould closed during the foaming stage. Thus, the injection machine required can be much simpler and consequently cheaper than the traditional injection moulding equipment. With thermoset foams such as polyurethanes made by dual component processing, injection pressures are very low, and mould tools need only resist pressure from the expanding foam. However, mould orientation and location of filling points are critical, especially in long and narrow cross-sectional area moulds (e.g. chair shells), to avoid foam density differentials within the length of the moulding.

Structural foams consist of a low-density cellular core sandwiched between solid integral skins. Foamed parts are mainly manufactured using either both types of injection moulding, or an extrusion process with a polymer melt or liquid monomers containing an inert gas. The inert gas can be generated in a number of ways, e.g. decomposition of a chemical by heat or water, boiling of a liquid due to the exotherm from reacting monomers, or by the addition of a compressed gas (usually nitrogen or fluorocarbon) into the melt before injection. The solution is injected into a mould cavity and is then allowed to expand in the cavity. With thermoplastics, a solid skin is formed by the solidification of the melt before expansion can take place. This skin forms an insulating layer for the core to expand by foaming. Experience with foams has shown that the stiffness and strength vary with the square of the foam density, for densities in the range $400-1\,000\,\mathrm{kg\,m}^{-3}$.

The advantage of a foam can be demonstrated by considering two components with the same mass, with the foamed version having the greater thickness. The bending stiffnesses can be compared for a simple rectangular beam, for solid polymer (density $1\,000\,\mathrm{kg\,m}^{-3}$) and foam (density $700\,\mathrm{kg\,m}^{-3}$). Calculation shows that the foamed beam is 43% stiffer than the solid beam with the same mass. However, this idealized situation is not fully correct because a moulding will never be homogeneous foam, but will have skins of solid polymer on the surfaces. This introduces more difficulties into the analysis of a foamed beam, although equations have been derived to solve the problem. Typically, polypropylene sections with thicknesses between 6 and 20 mm are recommended, and for design purposes the ratio of skin to core to skin is 1:4:1 respectively. These rules do not apply so strictly for the semi-flexible self-skinned polyurethane systems.

(ii) *Reinforced plastics.* Reinforcement can be obtained by adding fillers, glass beads or fibres to a basic polymer. Common fillers are clay, chalk, talc and wood flour which are added to the polymer in quantities up to 60% by weight. The effect of this addition is to improve rigidity, to reduce moulding distortion and to reduce costs, but the strength is frequently impaired and the melt flow length reduced.

Short-fibre composites are most commonly used in parts which are injection moulded or extruded. Predominantly glass fibres are employed, although carbon fibres and asbestos are occasionally used. The effectiveness of the reinforcement relies on the stress in a component being transferred from the polymer (normally termed the matrix) to the fibre. Thus, a good bond is necessary between fibre and matrix, and to this end the glass is coated with a coupling agent which is compatible with the matrix in which it is dispersed.

The mechanical properties of short-fibre composites are difficult to establish because the properties vary with the quantity of fibres and their orientation. For example, in a complex injection moulding the glass fibres may not be equally dispersed throughout the component, with some portions having little or no reinforcement. The greatest strength of a short-fibre reinforced component is in the direction of fibre orientation with the least strength perpendicular to the orientation of the fibre. The analysis of a component incorporating short fibres is difficult, and the mechanical properties in data sheets for short-fibre reinforced polymers should be used with considerable caution. Particular difficulties arise when a designer has to establish the performance of a component using information provided for uniaxially oriented test specimens. Great care must be taken in using this information and, if possible, data for randomly oriented specimens should be employed.

Careful attention must also be paid to the design of a mould, especially the positioning of gates, in order to provide optimum strength in the critically stressed areas. Furthermore, the shrinkage in a short glass-fibre-reinforced moulding is greatest perpendicular to the direction of orientation of the fibres. This is directly opposite to the shrinkage of unfilled polymers, and a mould designer must allow for this different shrinkage.

Long-fibre composites are mainly used with hand lay-up systems using polyester or epoxides as the matrix in which glass fibres are suspended. It is possible in long-fibre composites to calculate the strength and stiffness more accurately using conventional composite theory. For example, in a unidirectional continuous fibre composite, the equation for the total stress, σ_c, in the composite can be shown to be:

$$\sigma_c = \sigma_m V_m + \sigma_f V_f$$

where V is volume fraction, and the modulus E for the composite is

$$E_c = E_m V_m + E_f V_f$$

Thus the performance of the composite can be analysed.

2.3.4 Tolerance and dimensional control of products

The successful use of plastics requires an appreciation of the dimensional accuracy that can be achieved at reasonable cost. Although metal parts only rarely change after shaping or machining, plastics parts may be subject to relatively large dimensional changes associated with the following:

(i) Processing scatter: for mouldings, affected by uniformity of moulding material, machine setting and tool temperature
(ii) Condition of the equipment: for mouldings, manufacturing tolerances on tool dimensions, tool wear, and positional changes in movable parts of the tool
(iii) Operating and environmental factors affecting the product.

The dimensional instability of plastics might be regarded as intolerable by an engineer experienced only in metals behaviour, but the inherent flexibility and resilience of most plastics materials can be adapted by such design features as force- or clip-fits.

The shrinkage of thermoplastics which occurs on cooling after moulding may result in distortion: it is usually defined as the difference in dimensions between the (moulded) product after cooling and the (cold) metal mould. This shrinkage is caused by thermal contraction and by the relaxation of moulded-in stress resulting from orientation of the molecular chains. Thus the shrinkage of a material will be different perpendicular to the flow than in the direction of flow, because of the relaxation of the molecular stretching. For example, in a moulding using ABS, the shrinkage ranges between 0.4 to 0.7% in the direction of flow, and between 0.3 to 0.5% in the transverse direction. However, glass-reinforced materials are opposite to unfilled materials because the glass fibres inhibit the shrinkage in the direction of orientation/flow. Thus, for glass-filled thermoplastic polyester for a wall thickness of 2.5 mm (the shrinkage depends on the wall thickness), the shrinkage in the flow direction is 0.25–0.35%, and in the transverse direction 0.5–0.7%. Crystalline thermoplastics, such as polyethylene, polypropylene or polyamide 6.6 exhibit greater shrinkage from the melt (1.5–2.5%) than do amorphous plastics, such as polystyrene or poly(methyl methacrylate) (0.5–0.8%). Some guidance on achievable tolerances for plastics moulding is provided in BS 2026; 1953: BS 4042; 1966: DIN 16901; 1973: and by materials suppliers.

Dimensional tolerances of mouldings produced entirely within one mould part can be closer than for those produced across the parting line of the moulds and subject, therefore, to position variations and the thickness of 'flash'. Tolerances depend significantly on the type of material, shape and size of the component and degree of control of the process. Closer tolerances than those generally achieved, (usually monitored after at least 24 hours' conditioning) can be obtained by choosing materials exhibiting low shrinkage and low moisture absorption, and by processing them under strictly controlled conditions (with the processing equipment shielded from draughts, for example).

2.4 Importance of economics of processing in design

This important topic will be introduced by reference to the manufacture of a case for electrical equipment used in civil engineering (Figure 2.16). The cable avoidance tool (CAT) enables the underground track of power cables to be traced, and hence there are fewer accidents when excavating. Several produc-

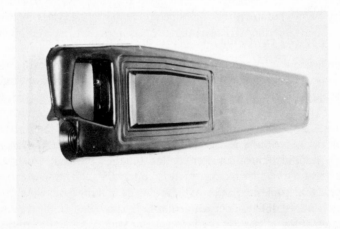

Figure 2.16 The Cable Avoidance Tool is made by Electrolocation Ltd., Bristol. It was designed by B.I.P. Ltd., Streetly, West Midlands, and is moulded by Brettel and Shaw Ltd., Oldbury

tion methods can be compared and the costs estimated for producing approximately 1 000 units per annum, the estimated market for CATs. This component can be shaped using either blow moulding or rotational moulding in one operation. Otherwise, it can be formed in two parts and assembled later, the parts being injection moulded, thermoformed or manufactured using hand lay-up or spray techniques. Thermoforming, in particular vacuum forming, lends itself to large simple open sections such as this. This size of product is limited to the size of sheet available, but the equipment and moulds are less expensive than those for injection and blow moulding. Fabricating articles in plastics is similar to sheet metal fabrication; it is a very labour-intensive and slow process.

Rotational moulding is suited to simple enclosed shapes like footballs, road markers and tanks up to 3 m^3 capacity. The equipment cost can vary widely, but moulds can be made cheaply and quickly from sheet steel.

The use of reaction injection moulding (RIM) of high modulus polyurethane offers a cheaper process relatively to screw-fed injection moulding. The basic dispense unit, with a couple of mould tools (aluminium or ideally inexpensive epoxide, for such low production runs), and clamping press, works out cheaper than for thermoplastic injection moulding equipment with a single steel tool. Some of the slower-reacting polyurethanes give longer moulding cycle time, but the dispense unit can be more fully utilized by feeding up to nine other moulds/presses in a production cell. Although polyurethane raw materials are twice the price of commodity thermoplastics, process energy requirements for dispensing are greatly reduced, the reaction and mould filling being carried out at or just above room temperature.

Table 2.5 shows the costing of these production methods. The costs of rotationally moulded and thermoformed parts are considerably lower than for other methods for an annual production of 1 000 units. It is also true that the costs will vary with the size of the components, the moulding cycle time and the complexity of a moulding. This costing also assumes that the equipment for manufacturing the rotational moulding is simple and cheap. However, such equipment may vary from very simple rotating parallel bars with gas jets beneath, costing less than £1 000, up to a very advanced automated rotational moulding equipment costing in excess of £100 000. The cheap system has a poor production rate (four mouldings per hour), is very labour-intensive and is suited to small production runs, whereas the more expensive equipment has a fast production rate (several mouldings per minute), is suited to lengthy production runs, is mainly automatic and its output is similar to injection moulding.

In general, there will be a most suitable production method depending on shape, size and quantity required for any component. For instance, if quantities less than ten are required, then it is cheap to fabricate from sheet. For quantities between 10 and 500, hand lay-up techniques are economical. The cost of producing a simple product weighing 200 g by different production

Table 2.5 Relative manufacturing costings for the production of CATs

Assumptions	Injection moulding	Reaction Injection moulding	Blow moulding	Thermo-forming	Rotational moulding	Hand lay-up	Fabrication
Cycle time	80 s[1]	5 min[1]	60 s	3 min[1]	1 h[2]	44 min[3]	3 h
Maximum throughput (units/hr)	45	12	60	20	2	1.36	0.33
Utilization (%)	65	65	65	65	65	65	65
Capital cost (£)	180 000[4]	50 000[5]	300 000[6]	12 000[7]	2000[8]	200	600
Installation charge (£)	18 000	10 000	30 000	600	1000	—	—
Original tooling cost (£)	40 000	7 500[9]	500	2000	2000	500	—
Total outlay (£)	238 000	67 500	340 000	13 000	5 000	500	600
Hours worked for 1000 units	34	128	26	77	770	1131	4615
Direct labour (@ £3.20/h)	1 person per 2 machines	1 person per 2 moulds	1 person per 2 machines	1 person	1 person	1 person	—
Setting (@ £3.70/h)	1 setter per 10 machines	1 setter per 40 moulds	1 setter per 10 machines	1 setter per 10 machines	—	—	—
Labouring (@ £2.90/h)	1 labourer per 10 machines	1 labourer per 40 moulds	1 labourer per 10 machines	1 labourer per 10 machines	1 labourer per 10 machines	1 labourer per 10 machines	—
Depreciation (% of capital)*	12,5	10	12,5	10	10	10	10
Machine tool overhaul (% of tool costs)	10[10]	10	10	10	15	20	—
Plant maintenance (% of capital)	4	4	4	4	4	—	—
Power rating (kW)	90	45	80	16	10	—	—
Consumables	—	£2.65[11]	—	—	—	£20	—
Joining of body (moulded sections)	£3 each	£5 each	—	£3 each	—	—	—
Final assembly cost	—	—	—	—	—	£3 each	£3 each

Annual cost for manufacturing 1000 Units (£):

Direct labour	54	102	42	246	2 464	3 619	14 786
Setting	13	12	10	28	—	—	—
Labouring	10	9	7	22	223	328	984
Supervision[1,2]	4	4	3	16	164	241	60
Depreciation	29 750	6 750	42 500	1 310	500	50	—
Machine tool overhaul	4 000	750	1 000	50	300	100	—
Plant maintenance	7 200	2 000	12 000	480	80	—	235
Power (@ 4.98p/kW)	157	72[13]	105	22	38	—	—
Consumables	—	2 650	—	—	—	2 000	—
Finishing costs	3 000	5 000	—	3 000	—	—	—
Total cost (£/1000 units)	44 188	17 349	55 667	5 174	3 769	6 338	16 047
Manufacturing cost (£/unit)[14]	44.188	17.349	55.667	5.174	3.769	6.338	16.047

Notes:

1. Two mouldings each.
2. Two moulds rotated simultaneously.
3. Minimum of one gel coat, and three layers of glass cloth/resin (@ 7 min per layer).
4. 500 tonne machine.
5. A RIM dispense unit normally will feed several moulds in separate presses, via a ring main and manifold system. Price here is based on equipment needed for two tools, each with a double cavity.
6. 30 litre capacity.
7. 1.5 × 1 m forming area.
8. Rotating drums over gas jets.
9. Estimates based on aluminium alloy tooling (5 000 to 25 000 units output). Cost will be reduced using epoxide tools for short run production, although their life will be less (250 to 1000 units).
10. One tool used after 10 mouldings.
11. RIM of PUs will include the cost of consumables, e.g. up to 25% of total moulding weight, release agent, tools and wash solvents.
12. One-fifteenth of direct labour.
13. RIM requires little energy for heating purposes, since it does not involve a melt process.
14. The polyurethane and glass-reinforced polyester materials will be $2\frac{1}{2}$ to 3 times the price of the LDPE materials used in the other processes. In some processes, the product will have to be made into two parts, which are then joined by an epoxide or urethane adhesive, or ultrasonic welding.

*Depreciation of capital equipment is based on a payback period of 8–10 years. However, current cost accountancy practice assumes payback in under 5 years; PU processing equipment is often estimated on the basis of a 2-year payback period.

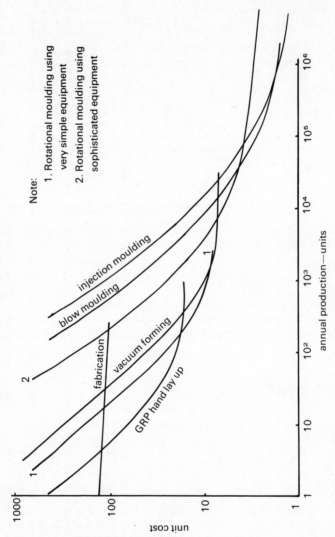

Figure 2.17 Comparative production costs for 200 g component

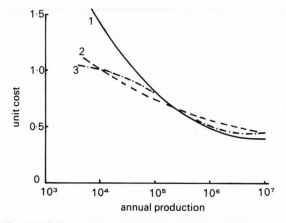

Figure 2.18 Comparative production costs for a 5-gallon drum
1. Injection moulding
2. Blow moulding
3. Rotational moulding

methods and for different production quantities is shown in Figure 2.17, but this can only be used as a guide, and it is important that designers should consider the product shape, size and production quantities before deciding on one method or another. In fact a cost-effective product can be produced if a cost analysis of several processes is conducted at this stage. Further details of cost versus production quantity are discussed by Johnson [2], and two examples are shown in Figures 2.18 and 2.19.

The annual production of mouldings required for the CAT indicated that the most effective manufacturing method would be either GRP hand lay-up, thermoforming or rotational casting, but experience had proved that GRP

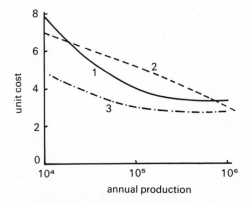

Figure 2.19 Comparative production costs for a 55-gallon lidded drum
1. Injection moulding
2. Blow moulding
3. Rotational moulding

was not satisfactory because the cases were, in fact, too expensive to produce. Thermoforming was possible, but this manufacturing method causes thinning at corners and in deep sections, and the mouldings would need to be joined around the circumference. Rotational moulding provides a sealed one-piece unit which is more robust than the thermoformed part.

For completeness of these comparisons, reaction injection moulding (RIM) has now been included in this new edition. Thus the case for RIM is developed on a hypothetical basis, particularly with the need to use an alternative polymer (low-density polyethylene being replaced by high-modulus polyurethane, three times more expensive, as the product material). In the USA, RIM and RRIM processes are used in direct competition to thermoplastic injection moulding in the automotive industry, for example for vehicle bumpers (fenders) at 5–20 kg shot size. To produce heavier plastics components, basic RIM dispense machinery will be cheaper than that for thermoplastics, (£110 000 compared to £300 000 with closed-loop control), and with cheaper tooling costs (£150 000 compared to £250 000). Until the early 1980s, cycle-times for polyurethane-RIM ranged from 3–10 minutes, (application of release agent contributing significantly), compared with 30–120 seconds for thermoplastics. To their advantage, polyurethanes are processed at near room temperature, saving on process energy requirements. The recently introduced polyurethane/polyurea systems have reaction rates which permit cycle times of 45–120 seconds, and with the promise of a new generation of polyurea-RIM, cycle times may be reduced to less than 30 seconds. Therefore, a decision that must be made by the designer, often in conjunction with the technologist, is whether production numbers, process costs and property requirements will justify the use of high-cost raw materials.

References

1. Rooke, D.P. and Cartwright, D.J. *Compendium of Stress Intensity Factors.* HMSO, London (1976).
2. Johnson, N.C. Cost comparison of rotational casting with injection moulding and blow moulding. *Brit. Plast.,* July 1974, p. 376.

Further reading

Ogorkiewicz, R.M. (ed.), *Engineering Properties of Thermoplastics,* John Wiley, New York (1970).
Powell, P.C., *Plastics for Industrial Designers,* Plastics Institute, London (1974).
Williams, J.G., *Stress Analysis of Polymers* (2nd edn.), Ellis Horwood, Chichester (1980).
Powell, P.C., *The Selection and Use of Thermoplastics,* (Engineering Design Guides No 19), Oxford University Press (1977).
Monk, J.F. (ed), *Thermosetting Plastics: Practical Moulding Technology,* Godwin [Longman Group] London (1981).
Ogorkiewicz, R.M. (ed.), *Thermoplastics: Properties and Design,* John Wiley, New York (1974).
Ehrenstein, G.W. and Erhard. G., *Designing with Plastics,* Hanser [John Wiley] (1984).
Morton-Jones, D.H. and Ellis, J.W., *Polymer Products; Design, Materials & Processing,* Chapman and Hall, New York (1986).
Powell, P. C., *Engineering with Polymers,* Chapman and Hall, New York (1983).

3 Styrene plastics

3.1 Polystyrene

During the 1939–45 war, supplies of natural rubber were denied to the Western Allies by the Japanese occupation of Malaysia and other areas producing natural rubber in the Far East. As a substitute, butadiene-styrene synthetic rubber was urgently manufactured on a large scale, requiring the manufacture of styrene on a considerable scale. At the end of the war, with natural rubber freely available, the demand for synthetic rubber was reduced, and the styrene plants were sold cheaply to the chemical industry. Styrene is prepared by the Friedel–Crafts reaction between benzene and ethylene to yield ethylbenzene, which is dehydrogenated to the monomer. It is not an important chemical, except as a constituent of polymers, which are produced by mass, solution or suspension polymerization, usually the last; emulsion polymerization gives polystyrene latex. The quality of the polymer is monitored by softening point, solution viscosity of a 2% solution in toluene, and the proportion of methanol-soluble material. The softening point is particularly affected by residual monomer, as is odour and the acceptability for food contact applications.

Although polystyrene (PS) is a commodity plastic, much of the recent growth has been in the modified grades; however, crystal polystyrene, (so-called because of its sparkling appearance, and not implying that it is crystalline), is available in a number of variants of different molecular weight. These include *general-purpose*—a compromise between high molecular weight for good mechanical properties and low molecular weight for adequate flow; *high molecular weight*—for better impact performance; *low volatiles*—increased softening temperature by up to 7°C; and *easy flow*—polymer of lower molecular weight, plus internal lubricant.

The commercial plastic is non-crystalline, with a softening point of approximately 100°C. The low softening temperature and the amorphous nature of the plastic mean that PS is one of the easiest plastics to mould, since temperature may be used to reduce the viscosity to acceptable levels. However, PS tends to degrade by depolymerization at temperatures above 150°C, so that long dwell times in the melt should be avoided. Further, due to its ready mouldability, there is a temptation to use long flow paths, and low mould temperatures, for economy in cycle time, resulting in high residual strain in the mouldings, with disastrous consequences on the strength. Other processing advantages accruing from the non-crystalline nature are lower mould shrinkage than for crystalline plastics, and a low heat content (enthalpy) to the processing temperature, again compared with crystalline plastics.

Since polystyrene is amorphous, it is transparent unless insoluble additives are used, and for the same reason it is soluble in aromatic solvents and chlorinated hydrocarbons, and has a strong tendency to stress-solvent crazing. As expected for a hydrocarbon polymer, PS has low water absorption and excellent electrical properties. The specific gravity is 1.07 and the plastic is hard and brittle; the strength is not exceptional, with pronounced stress-crazing prior to fracture. Polystyrene is susceptible to degradation by UV radiation (weathering); it is also flammable, burning with a yellow, sooty flame.

The properties militate against the use of this plastic in critical engineering applications, but PS is widely used in decorative and commodity packaging; examples are cosmetics and foodstuffs. An engineering use is approached in lighting fittings, where evenness and intensity of illumination are important, for instance in bookshops. There have been many trivial and transient applications resulting from the easy processing, rigidity, transparency and relatively low price.

Polystyrene has suitable rigidity and high-temperature performance to enable it to be formed into attractive light fittings, especially covers for strip lighting. However, the designer should be aware of the limited resistance of the material to UV radiation, which may limit its use with high-intensity fluorescent tubes. Furthermore, when PS is stressed, there is a tendency for the material to craze at stresses considerably lower than that for failure; crazing is visually unacceptable. However, PS is suitable in lighting fittings because of the low stress involved and the minimal impact abuse; nevertheless, sharp corners should be avoided and generous radii given to reduce the risk of premature failure.

For the highly competitive lower-priced cosmetics market, the packaging is often used to sell the product; thus it has to look attractive without increasing the selling price significantly. For an attractive product, the essential requirements are a good surface finish and colouration, whilst only a small amount of post-moulding distortion is acceptable; with the added property of transparency, PS is a reasonable choice. These same qualities are involved in the marketing of confectionery, but here it is the higher-quality end of the market which has adopted plastics packaging. This is preferred to the paperboard used at the lower end of the market: an example is shown in Figure 3.1. For economic reasons the melt temperature and the mould temperature are sometimes kept at low levels to obtain short cycle times: this leads, however, to stresses frozen into the moulding. These are detrimental to the mechanical properties and could result in premature failure. It is desirable, therefore, to mould with higher mould and melt temperatures, with conse-quently increased cycle times to reduce the moulded-in stress: a balance must be reached between cost and performance.

Figure 3.1 Crystal PS is used for the packaging of high-quality confectionery

3.2 High-impact polystyrene (HIPS)

The main deficiency of polystyrene is its brittleness; this may be improved by a
number of techniques, which are generally applicable in polymer technology:

  (i) *Increasing the molecular weight*: this inevitably increases the melt
      viscosity which, in turn, may result in higher residual orientation in
      mouldings and extrudates, with consequent deterioration in impact
      strength

 (ii) *Planar orientation*: this is applicable only to films, and, more recently,
      pipes and bottles

(iii) *Addition of plasticizer*: only a limited improvement in toughness is
      obtained, and there are significant losses in softening temperature,
      rigidity and strength

 (iv) *Copolymerization*

  (v) *Polymer blending*: a variety of blends is possible – homogeneous blends
      with other plastics of good impact strength, composites involving fillers,
      and blends with rubbers.

Although there is limited scope for improving impact performance by
increasing the molecular weight, (iv) and (v) provide the main methods by
which the toughness of polystyrene is improved.

  The addition of rubbers to PS is widely practised; the secret of obtaining
good impact behaviour without compromising rigidity is that the blend
should be heterogeneous. Butadiene-styrene rubber (SBR) may be blended
with PS, but this can lead to molecular mixing, resulting in plasticization and

little improvement in toughness. Better results are obtained if the rubber is dissolved in styrene monomer and the resulting syrup polymerized, leading to a dispersion of rubber particles in polystyrene. The greatest improvement in toughness is obtained if rubber latex (polybutadiene is preferred for good low-temperature properties) of particle size 0.1–1.0 μm is reacted to give a grafted coating, and the product then added to styrene monomer which is poly-merized. Usually 10–20% rubber is involved. The structure of this composite consists of a PS matrix, with inclusions of discrete rubber particles, some of which themselves contain smaller particles of polystyrene. The rubber phase is rendered visible in the electron microscope by staining with osmium tetroxide or iodine; the particles are too small to be seen by light microscopy, although they scatter light, leading to opacity.

The easy processing characteristics of polystyrene are retained in HIPS; indeed, for pre-gelled rubber particles, injection moulding is facilitated, but if they are larger than 10 μm, surface finish deteriorates. Thermoforming, especially vacuum forming of extruded sheet, is a favoured way of shaping simple objects, e.g. disposable beakers for drinks, and more complex engineering structures, such as refrigerator liners.

The modification of PS with rubber results in a considerable increase in impact toughness, by an order of magnitude, but with deterioration in other properties: 5–15°C in softening temperature and 20–40% in rigidity and strength (see Table 3.1). The ageing behaviour is also worsened by inclusion of rubber, but incorporation of antioxidant largely restores the situation.

These properties allow the use of HIPS in many applications, especially where impact abuse is encountered; however, the low softening temperature offers some limitation. Applications include refrigerator liners and fittings, toys, games and sports equipment, and radio and electrical equipment housings. As with PS, design with HIPS has been largely empirical, as in the following case study.

There have been several examples of the unsatisfactory use of PS, because of the brittle nature of the material; the application in toys is typical. Children frequently abuse their toys and thus a good impact performance is essential. General-purpose polystyrene (GPPS) breaks and exposes a sharp edge as an additional hazard. The advent of HIPS allowed the good features of PS to be retained, with the additional benefit of toughness at room temperature (the toughness falls dramatically below $-20°C$), for only a small increase in

Table 3.1 Properties of PS, HIPS and SAN

	PS	HIPS	SAN
Tensile strength (MPa)	34–51	27–43	65
Elongation (%)	4	20–50	2.5
Impact strength (J/m)	10–20	45–65	10.5
Vicat Softening Temperature (°C)	90	85	102

material cost. Several materials could be considered for the manufacture of toys: impact modified PP and HDPE are durable and appear superior to HIPS. However, HIPS is frequently preferred for the following reasons:

(i) PS can be easily joined using solvent adhesives
(ii) Amorphous materials have less shrinkage, and take more detail from the mould surface
(iii) Painting and plating are much easier for HIPS.

Many of these qualities of HIPS are implicit in the toy shown in Figure 3.2.

Refrigerator liners are simple open-section parts which could be made by a variety of shaping methods. However, the comparatively small numbers required for any one design precludes techniques such as injection moulding. Thermoforming is more favoured, but this restricts material choice; further, the material must stand abuse at 4°C, and withstand thermal cycling between this temperature and ambient conditions. Crystalline polymers are more difficult to handle since their melt strength is low. PVC thermoforms satisfactorily but has generally less suitable low-temperature properties.

The liner in Figure 3.3 is manufactured in sheet of 3 mm thickness which is heated on both sides and blown with compressed air. The part is subsequently shaped on a male former by vacuum. A minimum thickness of 1 mm is retained after moulding to ensure robustness, with the overall thickness 1.5–2 mm. When the liner is in position in the body of the refrigerator, the space is injected with polyurethane foam, giving rigidity in the protruding features of the liner and thermal insulation.

The low-temperature impact performance of HIPS is unsuitable for deep-freeze applications, where ABS is preferred; the penalty is that ABS is significantly more expensive than HIPS.

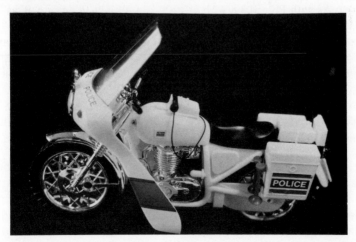

Figure 3.2 The Action Man Motorcycle manufactured by Kenner Parker (formerly Palitoy), Coalville, Leics

Figure 3.3 Refrigerator liner manufactured by Electrolux Ltd

As we have seen, HIPS is frequently in direct competition with grades of polypropylene.

3.3 Styrene copolymers

Although styrene has been copolymerized with a wide variety of comonomers, only the systems involving butadiene and acrylonitrile are of commercial importance; the former are employed as rubbers and, at higher styrene contents, as coating resins and rubber reinforcing agents. They will not be considered further. In the styrene–acrylonitrile system, one particular copolymer is unique, that containing 65–70% styrene (SAN), which is a preferred composition in respect of properties, and especially since it is easily manufactured. The reader is referred to other texts for details of copolymerization mechanisms, but it should be mentioned that it is rare to find copolymers formed at the same ratio as the comonomer feed; the copolymer quoted is such a case. Compositions containing more than 80% acrylonitrile have also been developed as barrier resins; they are melt processed but their progress is retarded by fear of the toxicity of residual acrylonitrile monomer.

Incorporation of acrylonitrile into a styrene polymer increases softening point and toughness, presumably by increasing the interchain attraction. For the same reason, resistance to non-polar solvents, such as fats and oils, is improved. Some data are given in Table 3.1. To summarize its properties, SAN has a softening point some 10°C higher than that of PS; is transparent, but with a tendency to yellowing, particularly on weathering; and is somewhat tougher and stronger than polystyrene. SAN is used for drinking tumblers, jugs and other transparent domestic ware, and for transparent and reasonably tough covers for domestic, electrical and automotive equipment, as shown in the following case study.

Many manufacturers of record players have supplied transparent covers for their equipment, so that record and mechanism are visible. A cover may be abused or dropped and so requires adequate impact strength. Further, it is an aesthetic part, so that distortion must be low, and preferably undetectable. For transparency, the choice is sensibly limited to amorphous polymers, which also have good after-moulding dimensional tolerances and low distortion. SAN was chosen as it has better impact properties than PS, retains the easy flow of PS, and is cheaper than PMMA. The characteristic yellow tint of SAN is usually masked by adding dye to produce an overall effect similar to smoked glass. The moulding is centre-fed into the underside of the top, but since the sprue leaves residual marks and flow lines, these have been disguised by positioning the decorative insignia.

Recently, a new family of styrene copolymers has been marketed, combining a high transparency (although marginally inferior to PS) with good impact properties and rigidity. They are block copolymers of styrene and butadiene, containing a preponderance of the former (70–80%); this constitutes the transparent matrix in which submicroscopic domains of (block) poly-butadiene are dispersed. These provide energy-dissipative mechanisms during impact, but are sufficiently small not to scatter light appreciably, although a 'blue haze' can be discerned. As with other block copolymers, processing and subsequent behaviour may not be straightforward. An example, a transparent component for a toy, is shown in Figure 3.4.

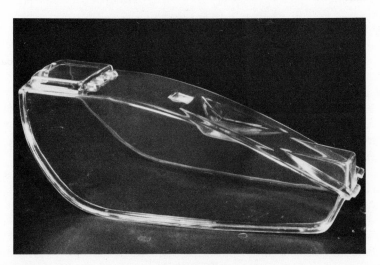

Figure 3.4 Transparent component for a toy manufactured by Kenner Parker (formerly Palitoy), Coalville, Leics

3.4 ABS plastics

The availability of SAN, of greater toughness and higher softening temperature than PS, gives an opportunity for composites analogous to HIPS. The
polar copolymer allows the use of polar butadiene-acrylonitrile rubber which,
pre-gelled in a blend at 20% content, gives a composition of very high impact
strength. A product with superior low temperature performance is obtained
from polybutadiene latex, grafted with a SAN-compatible coating on to which
the SAN is polymerized. Following much development work by the manufacturers, a wide selection of grades of ABS is marketed, sharing very great
resistance to impact abuse, with a softening temperature similar to that of PS.
ABS plastics are usually opaque, although a transparent version can be made
by matching the refractive indices of the resin and rubber phases. This is,
however, inferior to ABS and more expensive, so is not widely used.

The dominant property of ABS plastics is the very high impact strength,
which is coupled with less desirable features, such as a softening point similar
to that of PS, and comparatively low stiffness and strength for structural use.

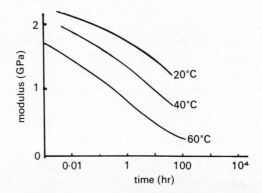

Figure 3.5 Tensile creep modulus of ABS

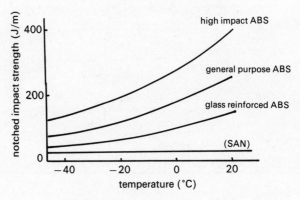

Figure 3.6 Impact properties of ABS

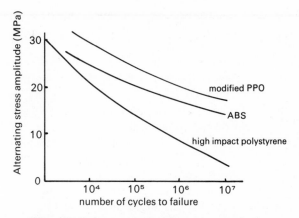

Figure 3.7 Fatigue properties of some styrene plastics

They retain, however, the excellent injection moulding behaviour of styrene plastics, particularly good melt flow, low mould shrinkage and excellent surface finish. Properties relevant to design are given in Figures 3.5–3.7. The rigidity can be improved by reinforcing fillers; glass fibre composites appear to be the most popular, although the surface finish and toughness are compromised. Attention is drawn to Figure 3.7, which records that the fatigue behaviour is not comparable with the excellent impact property.

Applications of ABS plastics reflect the processing behaviour and the high impact strength: vacuum cleaner covers, carpet sweeper housings, lawn mower housings, telephone handsets (replacing PMMA which in turn replaced PF), and hair-dryer housings of higher softening temperature than HIPS. They also

Figure 3.8 Flymo Hovermower™ cover

find application where good dimensional tolerance, coupled with good surface finish and adequate mechanical properties, is required: car instrument clusters, car fascias and trim.

Typical of the applications of ABS is the *Flymo Hovermower*™ cover, shown in Figure 3.8. This component, which replaces a similar unit in sheet metal, offers many advantages. It is aesthetically attractive, and does not require painting for protection. Impact blows, inflicted by missiles picked up by the rotating cutters, do not produce damage by fracture or denting. The unit is injection moulded and a good surface finish is an added attraction. All the fitting elements are included in the moulding, so no further fabrication is necessary, in contrast to its sheet-metal predecessor.

As a further case study, for many years the main body component of upright vacuum cleaners has been manufactured as a metal casting, since it was thought that metal was the only material able to withstand the abuse inflicted by the user. However, the cost of casting metal has increased dramatically in the last 25 years, and plastics have become increasingly attractive on economic grounds. The component is used to locate accurately several parts which interconnect, so that close tolerances are essential. A further requirement for this product is good visual appearance, since it is the part which customers examine most closely in the showroom. The choice of material depended mainly on impact performance. ABS is outstanding in this respect, and has suitable strength and good moulding properties at an economical price. Modified poly(phenylene oxide) – see later in this chapter – is also suitable. The cheaper ethylene and propylene polymers suffer from poor dimensional stability and unacceptable warping for this application (see Chapters 5 and 6). Polyamide 6.6 is a possibility, but the material is crystalline and, being sensitive to water content, will distort to a greater extent than other materials. Polycarbonate is very suitable, but is expensive.

The change from cast aluminium to ABS for this component required a complete redesign of the base to allow for the different properties of plastics, and the greater detail possible in the moulding operation. Several ribs were incorporated in the new design, including variations in wall thickness to withstand impact, short- and long-term loads. Such a design with varying section may lead to distortion and surface marks; the polymer is injected through several gates positioned in the central hole, thus preventing the appearance of surface marks which would result from a gate situated at the surface. Moulding the component to accurate dimensions required close control over the moulding conditions, under which circumstances a tolerance of 0.1% is possible, although normally, on parts of 75 mm in size a tolerance of 0.2–0.3% is more usual.

3.5 Polystyrene–poly(phenylene oxide) blends

A product range of increasing commercial importance is based on blends of PS with poly(phenylene oxide) (PPO). The latter is in fact the dimethyl

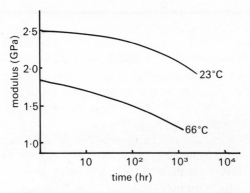

Figure 3.9 Tensile creep modulus of modified PPO

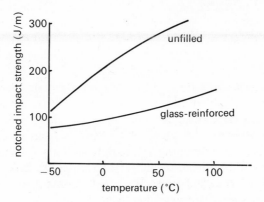

Figure 3.10 Impact properties of modified PPO

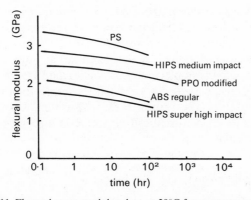

Figure 3.11 Flexural creep modulus data at 20°C for some styrene plastics

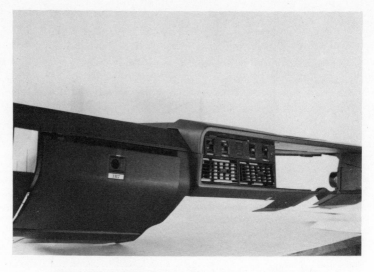

Figure 3.12 Triumph TR7 fascia moulded in modified PPO

homologue, which is used to restrict linkages in the polymer chain to the *para*-position. PPO homopolymer is a high-strength, reasonably tough, non-crystalline thermoplastic which is very difficult to process. It is also comparatively expensive. Somewhat surprisingly, PPO is miscible with PS on a molecular scale, and the blends are also capable of being further toughened by techniques analogous to those used in HIPS and ABS. Design data are given in Figures 3.9–3.11. Typically, grades with softening points some 30–40°C higher than polystyrene are available, with rigidity and strength approximately 50% higher than high-impact ABS, and with similarly enhanced long-term behaviour. The processing characteristics are inferior to those of ABS, particularly the melt flow, which is restricted; in consequence, it is usual to work at very high melt temperatures (280–300°C), and to use multiple gating. A property unusual in styrene plastics but exhibited by these blends is that they are self-extinguishing when ignited.

The blends have the good dimensional tolerance of amorphous plastics in general and of styrene-based plastics in particular, and are finding increasing use in car components, television set backplates, transformer cases, electrical equipment housings, etc. They can be made in structural foam grades, and as glass-fibre and mineral-filled composites. Further, the foams may be reinforced, giving a diverse family of materials, with many actual and potential uses.

Foamed modified PPO business machine components are large mouldings which need rigidity to retain several parts of the equipment (such as the keyboard), good dimensional stability and very good tolerances. Other performance requirements are good impact resistance, pleasing appearance and cost effectiveness. A structural foam based on a PPO–PS blend has been found very satisfactory at a density of $900\,kg\,m^{-3}$ with an integral skin.

As a concluding comment it must be pointed out that, in considering styrene-based plastics for particular applications, the poor solvent resistance might be a limitation. Simultaneous application of stress, in the presence of a solvent, may give premature failure due to environmental stress cracking. These comments apply equally to PS, HIPS, ABS and the PPO–PS blends.

4 Other amorphous thermoplastics

Amorphous thermoplastics were some of the earliest plastics, excepting polycarbonate, to find general acceptance; indeed, cellulose plastics were the first thermoplastics available commercially, and poly(methyl methacrylate), (PMMA), was a commercial product in the 1930s. Their continued application some 50 years later is a comment on their usefulness, and on their properties compared with those of polystyrene, since PMMA is almost twice the price of PS, and the cellulose plastics are significantly more expensive than PMMA. Polycarbonate, a development of the late 1950s, is some three times more expensive than PS, and so finds use in critical applications where performance rather than cost is the criterion of acceptability. There are other amorphous thermoplastics with yet more advantageous properties which have not reached the status of commodity materials; a selection of these with elevated service temperatures is reviewed in a later chapter.

In addition, one of the most versatile and widely used plastics, poly(vinyl chloride), (PVC), should also be included in this category. Since a small amount of crystallinity affects its behaviour, however, it will be considered separately in Chapter 8.

4.1 Poly(methyl methacrylate), (PMMA)

PMMA is available in two forms, sheet and moulding powder, which are manufactured by very different processes and which are shaped by different techniques. Sheet is frequently made by casting a syrup, consisting of preformed polymer is methyl methacrylate monomer, with added initiator, in glass cells, the polymerization being carried out over many hours. Alternatively, polymerization is carried out in an extruder, the sheet being shaped by an appropriate die on leaving the extruder. Whereas the former process can produce polymer of very high molecular weight, the latter method requires that the molecular weight be controlled to allow of die shaping. Moulding powders and general extrusion grades are made by suspension polymerization, involving molecular weight control by added chain transfer agent, often dodecyl mercaptan.

The cast sheet cannot be processed by methods generally applicable to thermoplastics; however, it can be thermoformed by vacuum, blowing or plug-assist methods. The deformation imparted is almost entirely *high elastic strain,* and the original sheet is regained if the shaping is heated above the softening temperature. This concerns the designer; he should be aware that at temperatures some way below the equilibrium softening temperature, the shaping will start to 'demould' or revert, as the residual stress lowers the effective softening temperature. Moulding powder or extrusion-grade polymer is processed conventionally, but has a somewhat higher melt viscosity than PS or the crystalline thermoplastics, and therefore requires more robust equipment. The viscosity is also more temperature-dependent than for most thermoplastics melts, necessitating good temperature control of the equipment. The plastic should be dry before processing, otherwise 'splash marking' may occur. In common with PS and all other amorphous thermoplastics, the mould shrinkage is low (0.8%).

The principal properties of PMMA are summarized below. The absorption of visible light is zero and the light scattering so low that for most practical purposes the transparency is perfect. (Signal attenuation is still significant, however, in the context of communications transmission lines.) Even with 'perfect' transparency, it should be remembered that reflective losses inevitably occur when light enters a material of different refractive index. PMMA is a material of high strength but is brittle, failing at low elongation. The very high-molecular-weight sheet polymer is stronger than the moulding grade, and significantly tougher. Data are given in Figure 4.1, which also records the crazing stresses for loading in air; the superior performance of the sheet material is again evident. This is also shown in the impact strength, which is some 50% higher than for the moulding grade, although the fracture is still brittle (Figure 4.2). Creep data are given in Figure 4.3. PMMA has a hard surface and is reasonably scratch-resistant; the surface hardness is amongst the highest for thermoplastics, 9H on the pencil scale. The softening temperature is slightly higher than that of PS at 105–120°C, being depressed by absorbed moisture and residual monomer. PMMA is very resistant to UV radiation and to weathering. This excellent behaviour may be impaired by residual stress in the shaping, which can add to the moisture absorption and desorption stresses

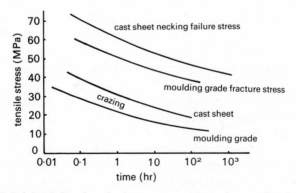

Figure 4.1 Poly(methyl methacrylate): rupture stress in tension as a function of time

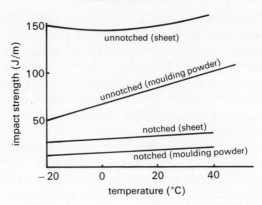

Figure 4.2 PMMA cast sheet and moulding powder: impact strength vs. temperature (notch tip radius 2 mm)

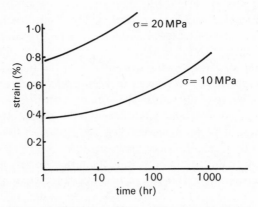

Figure 4.3 PMMA moulding powder: creep in tension

and lead to crazing. One of PMMA's least attractive properties is that it is readily flammable, with burning drips, restricting the use of the unmodified plastic in building applications; flame retardant grades have been developed, however, with improved performance in this respect. PMMA is resistant to salts, alkalis and dilute acids, detergent solutions, greases and oils, but many polar solvents, including alcohols, ketones and chlorinated hydrocarbons, are partial or complete solvents and promote crazing, particularly in stressed samples.

Applications are principally concerned with transparency and light transfer, with good weathering resistance, and with high surface hardness and durability. They include glazing in aircraft and other transport, transparent guards and covers, baths, wash-basins and sanitaryware, and rear-light assemblies for road transport. The last two applications exemplify the use of sheet and moulding grades respectively.

Producing large shapings frequently involves thermoforming techniques because large objects can be shaped easily and cheaply by this method. Amorphous plastics are more suited to this production method, and it is therefore predictable that acrylic sheet should be used in the manufacture of large thermoformed shapings such as baths. PMMA cast sheet has become established as the most suitable material because it has a higher softening point than PS or HIPS, and is rigid and hard, so that the surface does not scratch easily. Furthermore, it is cheaper than polycarbonate, does not absorb much water, and is unaffected by the numerous chemicals used in baths. A further advantage is gained by the use of cast sheet, because the strength is increased by the orientation in the direction of draw.

A rear-light assembly (Figure 4.4) consists of two main parts: the lamp body which contains the reflective pockets and retains the bulbs, and the cover, comprising several different colours in one moulding, or several components of different colours. The assembly is held together by self-tapping screws, which locate through the body and grip into bosses moulded into the lens (or vice versa). PMMA is used because of its good transparency, good colour retention, satisfactory chemical resistance and adequate impact performance at a moderate price. PS is cheaper but is very brittle, has poor solvent resistance, is attacked by petrol and tends to craze at comparatively low stresses. Polycarbonate is a very tough and rigid material with good transparency and, with the addition of UV stabilizers, is unaffected by weathering. However, it is more expensive than PMMA, and petrol may be a stress cracking hazard; polycarbonate is sometimes used in lenses requiring superior impact performance, and which generally do not come into contact with petrol (front signal and parking lamp lenses). Cellulose acetobutyrate is much tougher than PMMA, has good clarity, but is more expensive, has a lower working temperature, scratches easily and does not weather well. In the past, lenses were made separately, hot welded or dipped together, then

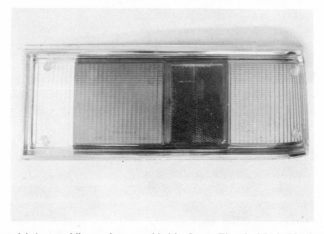

Figure 4.4 Automobile rear lens, moulded by Lucas Electrical Ltd., Birmingham

assembled to the base. This involved several assembly operations, and the unit was not always watertight. Currently, lenses are either moulded by two-colour machines (with two or more injection cylinders), or are moulded individually, then placed in another mould and encapsulated in clear polymer.

4.2 Cellulose plastics

Cellulose is a polymer which occurs abundantly in nature, constituting about one-third of the vegetable matter of the world. Cotton is almost pure cellulose, whereas wood contains about 50% of the polymer. Cotton linters and wood pulp are the usual raw materials for cellulose polymers. The former is digested at 120–150°C with dilute caustic soda, and bleached to give 99% pure cellulose, and wood pulp cellulose is extracted by a variety of chemical processes to a product of 88–90% purity, which can be enhanced by alkaline digestion and bleaching.

The empirical formula of cellulose is $C_6H_{10}O_5$, indicating its probable derivation from sugar (glucose, $C_6H_{12}O_6$ by elimination of water H_2O); indeed, the polymer can be hydrolysed to glucose. There is strong evidence that the structural repeat unit is cellobiose. This structure implies three hydroxyl groups per repeat unit (one primary and two secondary), with a consequent sensitivity to water, although cellulose does not dissolve, since it is crystalline and extensively hydrogen-bonded.

Exploitation of cellulose takes three routes: treatment to yield a tractable intermediate, with subsequent regeneration of the cellulose, and the manufacture of cellulose esters and ethers. The esters are the most important derivatives of cellulose, although cellulose itself is employed as a filler in cross-linked plastics, particularly urea–formaldehyde resins. Regenerated cellulose was the earliest man-made fibre and is still extensively used; in film form, cellophane was an early packaging film, now superseded by cheaper products with technically more desirable properties.

4.2.1 *Regenerated cellulose*

A number of processes are available, of which the most significant are the cuprammonium route and the viscose process; in the latter, cellulose is treated with caustic soda to give 'alkali cellulose', which is reacted with carbon disulphide to give 'cellulose xanthate':

$$\text{cellulose}-\text{O. Na} + CS_2 \rightarrow \text{cellulose}-\text{O}-\overset{\displaystyle \|}{\underset{\displaystyle S}{C}}-\text{S. Na}$$

cellulose xanthate

After forming into a film, the material is hydrolysed back to cellulose. Cellophane is very permeable to moisture, which is a shortcoming for a packaging film; the barrier properties may be improved by coating with poly(vinylidene chloride).

4.2.2 Cellulose esters

As we have seen above, the structural repeat unit of cellulose contains three hydroxyl groups which are available for chemical reaction, including esterification. As frequently found for reactions involving a crystalline polymer, homogeneous reaction usually requires that the polymer be in solution. The nitration of cellulose is, however, exceptional in that the reaction is progressive, the extent of nitration depending on the strength of nitric acid employed. Other esters of continuing commercial interest are cellulose acetate (CA), cellulose propionate (CP) and the mixed ester, cellulose acetobutyrate (CAB).

Cellulose dinitrate is obtained by treating cotton linters with a mixture of nitric acid, sulphuric acid and water (25:55:20), and treating the product with alcohol. Camphor, a favoured plasticizer, is mixed into the alcohol-wet cake, leading to the product 'celluloid', for which the most important application is the manufacture of table-tennis balls.

The esters derived from organic acids are more important than the nitrates, partly because they do not share the almost explosive flammability of the nitrates. Acetylation of cellulose reduces the capability of hydrogen bonding, decreases the polarity and increases chain separation; consequently esters with substantial substitution are amorphous. The acetates can be obtained at different levels of acetylation, but those at low levels are obtained by hydrolysis of the triacetate, not by direct esterification. The usual levels of reaction range from 2.2 for injection moulding grades to 3.0 for film and fibre applications. CA requires plasticizer to reduce the softening temperature and enable processing to be carried out without degradation: dimethyl phthalate and triphenyl phosphate are used in this way. Variation of plasticizer content, level of esterification and differences in the chain length of the parent cellulose lead to a family of plastics differing in softening temperature, hardness, strength and impact toughness. All grades are transparent, but are not particularly resistant to weathering.

Cellulose propionate (CP) and cellulose acetobutyrate (CAB) are more expensive than the acetate, but offer better dimensional stability (having lower water absorption) and improved toughness. For each ester class, a variety of grades is available, as with CA. Typical applications of CA, CP and CAB include transparent packaging, lamp shrouds and table lamps, control buttons, telephone dials, toys, tabular keys, spectacle frames and sunglasses, decorative trim for cars, and domestic appliances. Handles, from toothbrushes to cutlery and tools, and hammer heads, provide significant outlets for these plastics.

An important characteristic of cellulose acetate is its transparency; compared with other transparent materials it has much better impact properties

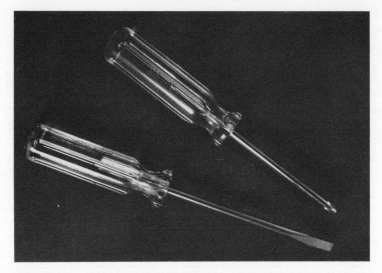

Figure 4.5 Cellulose acetate screwdriver handles

than PS, PMMA, SAN and P_4MP (see Chapter 6), but is inferior to polycarbonate, which is, however, more expensive. CA has been used for screwdriver handles for many years; these are extruded into the required profile, with ribbed or plain fluting, and cut to the desired length. They are then machined to produce domed ends or collars on the handles; trade marks, names, voltages and codes are hot-foiled or stamped afterwards. The production of tool handles by this method is cheaper than by injection moulding; the handles are produced in clear CA, so that they look attractive, and the tight fitting of the screwdriver shank can be seen (Figure 4.5).

4.2.3 Cellulose ethers

Although a number of cellulose ethers are known, the ethyl derivative is the only member finding plastics uses, mainly as a surface coating; others are water-soluble and are used in food processing. Commercial ethyl cellulose contains 2.15 to 2.6 ethyl groups per repeat unit and is obtained by treatment of alkali cellulose with ethyl chloride. It finds use in compositions for the strippable protection of metal parts.

4.3 Polycarbonate

Polycarbonates involve the simple carbonate linking group—CO_3— although only one such material, that based on bisphenol A (diphenylol propane), is important commercially. Structurally it is a polyester but, unlike aliphatic polyesters it is reasonably resistant to hydrolysis, particularly in the solid state, less so in the melt. Although polycarbonate (PC) is a linear polymer with a symmetrical structure, the melt-processed material is reluctant to crystallize, and PC articles and components are substantially amorphous and consequently transparent if unpigmented.

Grades of polycarbonate are available for injection moulding and for extrusion-based processes including blow moulding. As PC is a polyester and is susceptible to hydrolysis, it must be dried before processing, care being taken to keep the polymer dry in the hopper of the processing equipment. Compared with PS, the melt is processed at higher temperatures corresponding to its softening point 45°C higher. Further, the viscosity is less shear-dependent than for most thermoplastics. There is an even greater temptation than with polystyrene to process at a low temperature to reduce costs, but this leads to high residual orientation with possible catastrophic effects on the mechanical strength which otherwise, for properly designed and processed parts, can be outstandingly good. Gross frozen-in stresses may be identified by immersion of the mouldings in carbon tetrachloride, when such shapings develop extensive cracking; alternatively, the stresses may be monitored by birefringence measurements. Mould shrinkage is low (0.7%), and is reduced further by incorporating glass fibres; this may lead to anisotropy, due to orientation of the fibres in the flow direction, with consequent lower shrinkage (0.25%) compared with the transverse direction (0.4–0.5%).

The mechanical properties of PC have been of the greatest significance in its increasingly widespread application. Unusually for a rigid amorphous (glassy) plastic, polycarbonate is generally tough, as shown by a variety of impact data (Figure 4.6). However, great care must be exercised in the design of components in this plastic, since a wide range of factors can lead to embrittlement and premature failure; these include sharp cracks, (especially in thick sections), frozen-in stresses, cyclic stressing and a number of organic solvents which promote cracking. The deformation properties of PC are not particularly sensitive to temperature until the softening region above 140°C is reached;

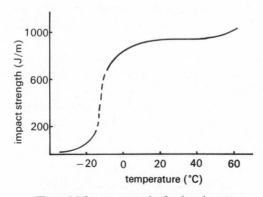

Figure 4.6 Impact strength of polycarbonate

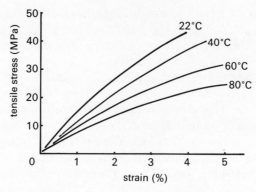

Figure 4.7 Polycarbonate: isochronous stress vs. strain curves. 1 000 hours

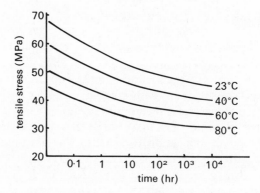

Figure 4.8 Polycarbonate: failure stress vs. time

isochronous stress vs. strain data for 1 000 hours loading are given in
Figure 4.7, and time-to-failure data portray a similar picture (Figure 4.8).

The other important attributes of polycarbonate are transparency (a quality
shared with all other unadulterated amorphous thermoplastics) and non-
flammability; the critical oxygen index is 25% for unfilled polymer. However,
with this index, it is likely that PC would continue to burn in a hot
environment. Also on the debit side, the chemical resistance is not outstanding,
and satisfactory weathering resistance is only achieved with additives.

The incorporation of fillers, particularly glass fibres, to enhance the rigidity
of PC is standard practice but, as might be expected, this has a deleterious
effect on the impact performance. Data illustrating the considerable improvement in modulus consequent on glass fibre reinforcement are given in
Figure 4.9 and Table 4.1; the deterioration in impact strength is seen in the
same table.

The range of uses of PC has been increased considerably in recent years,
many of the new applications not depending on transparency and being based

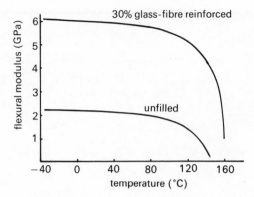

Figure 4.9 Flexural modulus of polycarbonate vs. temperature

Table 4.1 Effect of glass fibre on polycarbonate

Glass fibre content (%)	Flexural modulus (GPa)	Izod impact strength (J/m)
0	2.4	650 to 850
10	3.5	250
20	5.6	107
30	9.85	134

on composites, including foams. Blends of PC with other polymers are renowned for their excellent impact performance: these include the totally amorphous PC-ABS system and the partially crystalline blend with poly(butylene terephthalate). The latter will be discussed more fully in Chapter 7.

Examples of applications may be cited for which different selections of attributes are involved:

(i) *Toughness and transparency.* Bottles, ranging from feeding bottles to 25-litre containers (sterilizability is an added advantage, although prolonged boiling in water is detrimental to properties); protection shields or partitions in taxis or police vehicles; vandal-resistant light covers and illuminated signs; vandal-resistant roof lights; glazing likely to be subjected to impact loads (for instance in quarry buildings); inspection glasses in chemical plant.

(ii) *Toughness and high softening temperature.* Hair-dryer bodies; coffee makers; spotlight housings; housings for car light assemblies.

(iii) *Toughness and flame resistance.* Aircraft interior fittings; switch boxes for mines and quarries; high voltage plugs and sockets.

(iv) *Stable and consistent electrical properties.* These enable PC to be used particularly under extreme environmental conditions in small electric motors, casings for power tools and steam iron designs.

Figure 4.10 Polycarbonate component of vacuum cleaner manufactured by Electrolux Ltd.

An Electrolux suction cleaner (model Z345) has the motor and fan assembly suspended between a rubber mounting and a bracket (Figure 4.10). The motor assembly is forced into the rubber mounting and isolated from the bracket by a spring. Thus the bracket is used to compress the spring, and is loaded continuously throughout its life; furthermore, the cable feed mechanism locates into the bracket, and two electric contact rings are incorporated to enable the electricity supply to be connected between the rotating feed mechanism and the stationary bracket. The electrical power can then be connected to the motor. This component (Figure 4.10) requires good resistance to creep, especially at localized temperatures above 50–60°C, good electrical properties, and good impact behaviour to reduce the possibility of fracture if the cleaner is dropped accidentally. The dimensional requirements are also critical, to within 0.15%.

Previously phenolic mouldings were used for a similar application in a much simpler unit, but were not considered suitable in the present context because of the detail required. Moulding grades of PF have poorer flow properties than thermoplastics and, as a result, are more difficult to form into deep thin sections or items with many complex features.

Many thermoplastics can be considered for this application; although glass fibre reinforcement provides much greater rigidity, improved strength and a higher heat distortion temperature, these materials are more brittle than the unfilled ductile polymers and have unsatisfactory impact behaviour for this application. The high-temperature creep properties of ABS are unsuitable, and polyamides not only absorb water, which affects properties, but also have poor dimensional tolerances. The three most suitable materials are therefore modified PPO, PC and polyacetal (see Chapter 7); a comparison of their properties indicates that PC is the preferred material. Furthermore, its transparency helps to ensure that the electrical riders make contact with the two pick-up rings fixed to the brackets when assembling the unit.

The bracket has been designed with a constant section thickness of 2.5 mm to ensure that the component has uniform shrinkage and negligible distortion. It incorporates several radial ribs to provide a rigid and accurate location for the motor. The toughness of polycarbonate enables two threaded inserts to be positioned after moulding. Finally, the design incorporates retention lugs for the electrical cable and features to retain an on-off switch; it also isolates the electrical parts from the metal wind-up spring.

5 Propylene plastics

Propylene plastics have consistently enjoyed a very rapid growth both nationally and internationally; there seems to be little indication that they have reached their limits. The UK usage (Table 5.1) illustrates this growth, which is founded on the versatility of propylene plastics.

World consumption is divided almost equally between Western Europe, Japan and America; the total tonnage is exceeded only by polyethylene and PVC.

The monomer, propylene, is obtained by the thermal cracking of naphtha, (crude oil light distillate); ethylene, propylene and higher homologues are separated by fractional distillation at low temperatures. In the early 1950s Ziegler successfully developed a complex catalyst for the polymerization of ethylene. Natta applied Ziegler catalysts to propylene and obtained a crystalline polymer of high molecular weight. From Natta's discovery in 1954, it took less than four years to produce development quantities of polypropylene, in spite of the hazards associated with the catalyst and with the handling of explosive dusts. (Polymer fines are very explosive: a suspension of $1\ \mu m$ polypropylene particles in air can be detonated by a 2 mJ electrical discharge.)

5.1 Homopolymer and impact modified grades

The homopolymer has a crystalline melting temperature of 168–170°C; a by-product is the non-crystallizable variant, atactic polypropylene, (contaminated with low-molecular-weight crystalline polymer), which has little commercial value. This was inevitably isolated in the diluent-based polymerization, but recent manufacture, with more discriminating catalysts, seeks to exploit the whole product. The early polypropylene was of very high molecular weight and was almost impossible to process; there was even

Table 5.1 Use of propylene plastics in the UK

Year	Usage (tonnes)
1984	290 000
1985	330 000
1986	370 000

difficulty in measuring its melt viscosity, which caused a reappraisal of the Melt Flow Index (MFI) test procedure. The early product was degraded thermally at 260–280°C to give products which could be processed by conventional methods. Oxidative degradation was possible at much lower temperatures, but gave odorous products. Thermal degradation narrowed the molecular weight distribution, yielding products of good toughness for a given melt viscosity. However, polymerization followed by degradation, apart from giving a product with an unacceptable smell, was not attractive economically. There has been renewed interest recently in the balance of properties achieved in thermally degraded polypropylene; in particular the compromise between ease of processing and mechanical properties can be pitched at a different level, giving both improved productivity and better performance. For fibre applications these 'controlled rheology' (CR) grades have advantage in lower spinning temperature for optimum strength. The degradation can be achieved by mechanical or thermal treatment, γ-radiation, oxidation, or by the addition of peroxides. The last method is the most widely used production technique; the main characteristic of the CR polymer is a narrowed molecular weight distribution.

Control of molecular weight at the polymerization stage was achieved by adding hydrogen as transfer agent, or modifier, but the product has a wide molecular weight distribution, and is comparatively brittle for its melt viscosity. Thus, brittleness was a limitation to the development of polypropylene, particularly in mouldings; other problems encountered were oxidation in processing and use, poor resistance to UV-induced degradation and oxidation, and the development of structure (crystallinity) in the melt if low temperatures were employed to minimize degradation. Instability was cured by additives, giving particular fillip to oxidation studies and the development of new antioxidants. More recently, the use of UV-stable antioxidants has given PP compounds with still better weathering resistance. The ability to

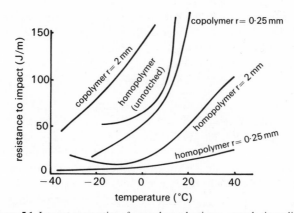

Figure 5.1 Impact properties of propylene plastics r = notch tip radius

process at high temperatures has eliminated problems of melt structure.

The relative brittleness of the homopolymer can be seen in Figure 5.1; the material, although ductile at 20°C, is embrittled either by a small reduction in temperature or by a slight increase in the severity of stress concentration. As with polystyrene, the toughness may be improved in a variety of ways: the most applicable are discussed below.

(i) *Addition of rubber*, originally butyl rubber, but now preferably ethylene-propylene or ethylene-propylene-diene rubber, is an effective method but necessarily involves a separate processing operation. Such products have been developed for car steering wheel covers, and for car bumpers (fenders). The apparently similar procedure of blending with atactic PP does not give a significant improvement in toughness, since atactic PP is a very poor rubber, having its glass transition not far below room temperature.

(ii) *Random copolymerization* with a second monomer, e.g. ethylene, results in improved toughness but at the expense of much reduced rigidity, an inevitable consequence, since the crystallinity is reduced by the inclusion of the second monomer. The melting temperature is also reduced significantly, for example, a 97:3, propylene–ethylene random copolymer had a melting point of 150°C. Such copolymers are manufactured commercially and used for film (for bonding to PP textiles, e.g. sacks), for bottles and for heat-seal coating, applications where the reduced melting point is not a serious shortcoming and may indeed be an advantage. Film based on these copolymers is replacing cellophane in the packaging of cigarettes.

(iii) *Block copolymer* of propylene and ethylene by sequential polymerization. This is the generally preferred solution to the problem of brittleness. Although frequently designated as 'block copolymers', very little block copolymer is formed in fact, although crystallinity of the two homopolymer species can be detected. The important material, however, appears to be ethylene–propylene rubber, possibly attached to a polypropylene or polyethylene chain. The improvement in impact behaviour is achieved without too serious a deterioration in the rigidity, which is reduced by 10–15%. The melting temperature is also reduced by a few degrees, compared with homopolymer (160–165°C compared with 168–170°C).

A substantial proportion of the PP now manufactured is in impact-modified grades which, in common with the homopolymer, show considerable tendency of the melt to supercool. This is illustrated in Figure 5.2, where the difference between the melting temperature, curve (*a*), and the crystallization temperature, curve (*b*), is some 70°C, even with a cooling rate as low as $16°C\,min^{-1}$. The importance of cooling rate can be seen in Table 5.2; the crystallization temperature for the rates of cooling inherent in commercial shaping processes is likely to be about 90°C, unless nucleated grades are used.

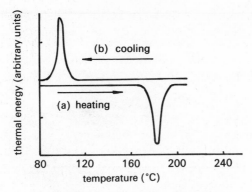

Figure 5.2 (*a*) Heating curve for polypropylene
(*b*) Cooling curve for polypropylene

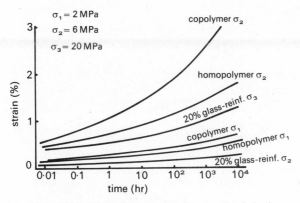

Figure 5.3 20°C tensile creep data for propylene plastics

Table 5.2 Crystallization and melting of sequential propylene–ethylene copolymers

	Crystallization temperature (°C)			Melting temperature (°C)		
Sample cooled (°C min^{-1})	0.5	5.0	50	0.5	5.0	50
Diluent	119	109	98	161	162	161
Gas phase	124	113	98	164	163	162
Diluent +0.25% talc		119			162	

Polypropylene is reasonably rigid with good creep resistance (Figures 5.3, 5.4). Further, the material has adequate load-bearing capacity for many applications (relevant data are given in Figure 5.5); the suggested strain limit is 3% (1% for welded components). Although the impact behaviour of the homopolymer is barely acceptable, that of the modified grades is excellent, as

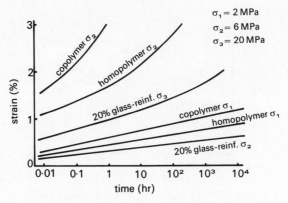

Figure 5.4 60°C tensile creep data for propylene plastics

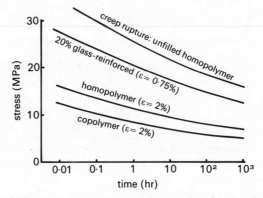

Figure 5.5 20°C creep rupture and isometric stress data for propylene plastics

illustrated in Figure 5.1. All grades of PP are capable of forming an 'integral hinge', a thin web of polymer of high strength and durability to flexing.

A particular application, the marine battery lid, is treated in greater detail below. Propylene-based plastics exhibit the chemical inertness associated with paraffin hydrocarbons, of which class they are high-molecular-weight members; this allows their use in contact with a wide variety of liquids, both domestic and industrial. In view of their structure, propylene polymers have excellent electrical properties, including very low dielectric losses.

On unusual feature of the structure–properties relationship is the existence of a number of crystalline forms of PP. This might have remained a scientific curiosity, except that some dyestuffs used for colouring the plastic nucleate the unusual crystalline form, the β-form, which is associated with type III spherulites, and inferior mechanical properties. Some batches of PP, particularly propylene–ethylene sequential copolymers, have a propensity to form

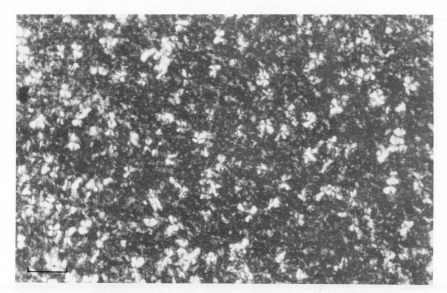

Figure 5.6 Micrograph showing an unusually high concentration of Type III spherulites. Scale bar 100 μm

type III spherulites; gas-phase polymers sometimes show a surprising concentration of these species (see Figure 5.6). It may be more than coincidence that such products appear to be relatively brittle. This form is also favoured by shear during crystallization, and thus may occur during welding.

Polypropylene has for many years been used for car battery cases, because of its light weight, high impact strength, chemical resistance and fast moulding capabilities compared with other plastics. It is no surprise that lids for such batteries should also be made from PP. AB Tudor (Sweden), however, designed other functions into the lid. Marine batteries are frequently carried between ship and shore and, therefore, require handles. An immediate solution would be to fit two 'clip-on' handles to the lid, which would be completed by fitting another moulding to seal the filler holes for each cell. The component would be fabricated from several parts; however, real economies would result from making the lid of a single moulding, incorporating integral hinges. In the final design, two integral hinges are employed for each of the two carrying handles (Figure 5.7a). When folded down, each handle presents a flat upper surface, flush with the filler caps and is retained in that position by snap-fitting two square hollow projections on the handle over circular studs moulded in the base of the lid. A further function of the handles is to cover the terminals when in the closed position. In the raised position, the handles are brought together and the weight of the battery is shared between the four hinges. Furthermore, at each end of the lid is a flap bearing three filler caps, and integrally hinged to the lid. When the flaps are in the closed position, each cap seals the aperture for its cell (Figure 5.7b). Integral hinges can be injection moulded, heat formed or

Figure 5.7(*a*) Impact-modified PP marine battery case; carrying handles in use

Figure 5.7(*b*) Impact-modified PP marine battery case: cell cap seals. The case of the Tudor marine battery is injection moulded by Boras Konstahts AB (Sweden) and the lid by Norsk Teknisk Parsalens AB (Norway). (*a*) and (*b*) by courtesy of ICI Ltd.

Figure 5.8 Container formed of two blow-moulded parts joined together by integral hinge. By permission of RB Blowmoulders Ltd.

machined; they derive their strength from the molecular orientation which is initiated during the moulding process, and completed by subsequent flexing of the hinge. The normal range of hinge thickness is from 0.25 to 0.6 mm, depending on the stiffness of operation and the strength or tear resistance required.

Another application exploiting an integral hinge is shown in Figure 5.8, which is a central fitting for Ford heavy lorries: the two halves of the container are blow-moulded simultaneously with the formation of the integral hinge.

The pattern of usage of polypropylene is dictated by the physical and chemical properties and by price, particularly the price in relation to competitive materials and processes. Both the homopolymer and modified variants are available in grades to suit a wide range of conventional shaping processes; in addition, special grades have been developed for novel processes, such as the 'solid-phase pressure-forming process' for shaping PP into thin-walled containers at temperatures which, although elevated, are below the melting temperature of polypropylene. This overcomes the problem of inadequate melt rigidity, which restricts conventional thermoforming.

The properties detailed above lead to the following main areas of application.

(i) *Homopolymer*: biaxially oriented film and unoriented cast film for packaging and electrical capacitors etc; uniaxially oriented film and fibrillated film for carpet-backing, sacks, twines, ropes etc; uniaxially oriented strapping tapes; spun fibres; injection-moulded measuring cylinder for photographic chemicals; blow-moulded bottles and containers.

(ii) *Impact-modified grades*: injection mouldings, e.g. bottle crates, tote boxes; chairs; pipes; liquid storage tanks, etc. The particular cases of a car coolant reservoir tank (Figure 5.9) and a chair shell (Figure 5.10), exemplify uses of these materials.

Car radiator reservoir tanks are now made in propylene plastics, because they offer advantages over metal in design flexibility, light weight and no corrosion. The containers are blow-moulded for the speedy production of large quantities, as opposed to rotational casting, and complicated shapes can easily be moulded (Figure 5.9). Service requirements are to withstand under-bonnet temperatures; to be resistant to anti-freeze and brake fluids; the translucency, enabling the coolant level to be seen, is an added advantage. Heat-stabilized, impact modified PP satisfies the performance criteria, and provides a tough and robust part. However, blow-moulding is limited in the number of inlets/outlets possible; frequently, additional outlets are added by heat-welding. The filler necks are injection-moulded in glass fibre-reinforced PP to provide a more rigid section to locate the traditional radiator cap. The manufacture of these tanks has seen a number of variants of shaping techniques and materials. Some are injection-moulded/heat-welded compo-

Figure 5.9 Coolant expansion tank for Jaguar car, blow-moulded in propylene-ethylene sequential copolymer. By permission of RB Blowmoulders Ltd.

Figure 5.10 Pel stacking chair moulded in impact-modified PP

nents in natural or glass-fibre-reinforced PP. Unfortunately, injection-moulding demands an easier flow grade of polymer compared with blow-moulding (or mouldings with high residual strain are produced, and this unfavourably affects the compromise between processing and environmental stress cracking). Ethylene glycol, the main ingredient of anti-freeze, is a significant cracking agent for PP; the position is exacerbated further by the large area of welding, which is usually far from perfectly carried out.

The design of chairs is continually scrutinized; the result has been a multitude of chair designs utilizing most of the commodity plastics. Many of the designs were for specialist uses, and have been of little commercial significance. However, since the 1930s there has been a market for low-priced seating for canteens, assembly halls and general use. This market required that these chairs could be stacked together to be stored compactly, and was obviously influenced by the large market for mass-produced chairs, and the commercial importance of this type of chair (Figure 5.10).

The major factors in the specification required the chair to contain a main structural element to support the back-rest and the seat; the material should be sufficiently flexible to be comfortable in use and the product must be attractively priced. The two most appropriate materials are high-density polyethylene (see Chapter 6) and polypropylene. HDPE has better low-temperature impact properties which could be an advantage for external seating in cold climates. However, when PP was subjected to an accelerated user test to simulate the load exerted by a 100-kg person pressing backwards on the chair for 100 000 cycles, the seat retained a permanent set of only 3.17 mm at the top of the backrest. This convinced the designers that impact-modified PP was suitable as a chair shell, as well as possessing superior creep properties to high-density polyethylene. A special grade of copolymer was chosen which possessed good flow characteristics, rigidity, gloss and good impact properties. An antistatic additive was also incorporated to reduce the amount of dust attraction and most recently, the fire retardancy has been improved markedly.

Comment has already been made of the use of PP which has been toughened with rubber, usually EPDM, in applications such as car bumpers (fenders). The application of surface coatings to modified PP mouldings used as external components on cars, e.g. front spoilers, is problematic, especially with long-term interfacial bonding. The theme of employing propylene polymers as reinforcing agents in a variety of rubbers, including thermoplastic elastomers, is being exploited increasingly. Additional advantages accrue from cross-linking the rubber in the presence of PP, when undoubtedly some interlinking occurs, which stabilizes the dispersion of PP and the rubber. These materials, exemplified by Santoprene™ (Monsanto) and Kraton G™ (Shell International), are penetrating markets such as seals and gaskets, in addition to being already accepted in shoe components.

5.2 Filled polypropylene

One of the main application deficiencies of PP is its inadequate rigidity, whilst the high mould shrinkage is frequently a problem in processing. Both features can be improved by the use of composites (blends of PP with fillers), which may also reduce the price of the material. Composites involving the addition of up to 30% calcium carbonate have been used for wastepaper baskets, where the increased rigidity is required in a cheapened product and the reduced strength is not a serious problem. For more exacting applications, structured fillers such as talc, mica and asbestos, are used, the first being widely and increasingly employed (data are given in Table 5.3).

A typical application for talc-filled PP is heater boxes for many cars and trucks, the use of this particular material allowing injection moulding of a complex shape of adequate stiffness, and with softening behaviour which allows its use under the bonnet. Traditionally, heater boxes were fabricated using phenolic cross-linked plastics; now they are made from thermoplastics, because of the design freedom and the ease of manufacture of complex parts. Unfilled PP suffers from poor dimensional tolerance and warping, and is considerably affected by temperature; on the other hand, talc-filled polypropylene has superior after-moulding tolerances and less warpage, because the filler reduces the mould shrinkage. Talc-filled grades are recommended for applications demanding better dimensional stability, higher rigidity and improved heat deflection temperature, and cheapness. However, the addition of talc is detrimental to other properties, the impact strength is reduced, the flow length is reduced and the specific gravity is increased. This example of a heater box is a complex moulding: several side cores are used to form features which are not parallel to the direction of mould opening. Talc-filling also reduces sinkage on the surface of the moulding, enabling thick sections to be joined, with only a faint trace of a sink mark. A heater box is illustrated in Figure 5.11.

For the best reinforcement, glass fibres are used, but it is essential to provide a good bond between reinforcement and matrix; composites involving carbon fibres are of little commercial interest for PP. Relevant design data for glass-fibre-reinforced PP are given in Figure 5.5. Examples of the use of glass-reinforced PP include air filters, lamp housings and the filler neck of car coolant-reservoir tanks. Some coolant reservoir tanks have been manu-

Table 5.3 Properties of talc-filled polypropylene

Property	Unfilled copolymer	40% Talc-filled copolymer
Flexural modulus (GPa)	1.38	3.55
Izod Impact (J/m)	133	48

Figure 5.11 Heater box from a Ford vehicle

factured in glass-reinforced PP, although ethylene glycol at high temperatures is not the most sympathetic treatment for the comparatively low-molecular-weight polymer inherent in reinforced grades. In the design of components in glass-fibre-filled PP, it should be remembered that injection-moulding will induce orientation in the direction of flow, to the detriment of properties in the transverse direction.

5.3 Foamed polypropylene

Another way of improving the economics of polypropylene usage is in foams; structural foam applications are of increasing importance, and agricultural produce containers of half-tonne capacity are already in use. Foams are being used in place of solid mouldings to give enhanced stiffness and greater economy: an example is the handle of a light-duty chain-saw where rigidity, strength, attractive finish and economy are important. Reinforcing fillers may be incorporated in the foam; one of the largest mouldings based on glass-reinforced PP is the static drum of the Philips washing machine. The modern unit comprises few major parts, one of which is the tank, which is used to locate the rotating drum and the motor, as well as inlet and outlet pipes, suspension

Figure 5.12 Philips washing machine tanks—(left) vitreous enamelled steel and (right) coupled glass-reinforced structural foam PP, moulded by Cabinet Industries Ltd., Alfreton. Photograph by courtesy of ICI Ltd.

springs, balance weights and damper mountings. In the past, front-loading machines contained vitreous enamelled steel tanks, which are very expensive (Figure 5.12a). The advantages that plastics immediately offer are good corrosion resistance, even to the most aggressive detergents, and low cost, since the component can be made easily in one injection moulding. The only doubts about this product were the strength and stiffness of the plastics materials, and in the very high tool costs incurred for moulds with several sliding cores and of large size. The material chosen was coupled glass-reinforced structural foam PP, since it has good chemical resistance, and the glass fibres increase the modulus. Foaming enables the product to adopt rigid thick sections which are dimensionally consistent, and the component will withstand thick and thin sections without undue distortion (Figure 5.12).

Foaming is produced by physical or chemical blowing agents, or by introducing nitrogen gas into the melt at the nozzle at the time of injection. Whichever method is adopted, when the melt is injected into the cavity, the material foams, expands and fills the cavity. The melt forms a solid skin which has been in contact with the mould surface and has solidified, giving an outer solid skin and an inner foamed core. The skin/core ratio depends on the thickness of the moulding, and the density of the core depends on the amount of material injected into the cavity. In structural applications it is not advisable to use a core density of less than $700 \, \text{kg m}^{-3}$ for unfilled polypropylene, and the wall thickness should be not less than 6 mm, otherwise no foaming will occur.

The advantage of a foamed structure arises when a part is subjected to flexural loading, for which case the stiffness is proportional to the cube of the depth of the beam. A further useful feature of foaming, compellingly demonstrated by the hull of the Topper sailing dinghy, is that mouldings of large area can be made without the need to employ impossibly large locking or clamping forces. The huge amount of melt for this moulding (20 kg) requires that two injection presses feed the mould in parallel, but the locking force for the two mouldings for the hull (of total projected area $3.5\,m^2$) is only 3 200 tonnes and is entirely feasible. The absence of sink marks in this largest injection moulding in commercial production is an added advantage: unreinforced impact-modified PP is the basis for this foamed structure.

Another way of creating a foam is to incorporate hollow sphere fillers; such composites, known as *syntactic* foams, have a superior surface finish compared with conventional foams. Flammability can be reduced by additives, e.g. antimony trioxide plus chlorinated waxes; such additives, while retarding the ignition, lead to increased smoke emission. The flammability of PP is a hazard restricting its widespread use in the interior of buildings.

5.4 Miscellaneous applications of propylene polymers

Typical of new applications in the packaging field is an extruded card substitute based on talc-filled sequential copolymer which can be extruded and cut, creased, folded and glued like cartonboard, and which handles well on high-speed carton-making machines. Alternatively, it can be thermoformed. The product is capable of withstanding deep-freeze conditions, is free from water absorption and is resistant to oils and greases. It is ideally suited as a container for microwave oven usage and, as it is inherently stronger than cartonboard, thinner sheets can be employed.

Corrugated sheet extruded directly from polymer is a strong competitor to corrugated board on cost and particularly on technical performance; the new product is lighter in weight and is unaffected by water, oils and solvents.

Oriented blow-moulding technology has been applied to polypropylene bottles: through biaxial orientation, PP bottles gain significantly in clarity, as well as in rigidity, permitting thinner walls. Another advantage for PP bottles is the high heat distortion temperature, which allows the container to be capped after hot-filling.

Although electroplating grades of PP are available, the important development in plating appears to be a pretreatment process which eliminates the need for special grades. It has the further advantage of requiring fewer steps than other methods, and hence is low in cost. A reactive chemical is used to swell the surface of the PP; evaporation of the solvent leaves the reactive chemical trapped in the surface, which is then brought into contact with an aqueous salt solution, making the surface electrically conductive. The pretreated surface can then be plated by standard methods. As well as operating at lower temperatures, chromic acid and expensive palladium chemicals are not used, making the new pretreatment very competitive with alternative methods.

6 Other polyolefin plastics

In addition to propylene polymers, the following polyolefins are produced commercially.

(1) *Polyethylene*
- (i) High-density (HDPE), produced by comparatively low-pressure routes with Ziegler or complex oxide catalyst
- (ii) Low-density (LDPE), traditionally made by a high-pressure route
- (iii) Linear low-density (LLDPE), comparable in structure and general properties with LDPE and manufactured by copolymerizing ethylene with butene, hexene, octene or 4-methyl pentene at low pressure
- (iv) Very low-density (VLDPE) is an extreme version of LLDPE
- (v) Ethylene copolymers, typically with polar monomers such as vinyl acetate, are made by LDPE technology
- (vi) Blends of various types of ethylene polymers are now widely used; sometimes their use is deliberate, sometimes incidental.

(2) *Polybutene*
This is an isotactic polymer, produced by Ziegler–Natta catalyst by a process analogous to that used to manufacture polypropylene.

(3) *Poly(4-methyl pentene)*
Another isotactic polymer produced by Ziegler–Natta catalyst.

6.1 Polyethylene

Chronologically, the manufacture of LDPE preceded that of HDPE, but since the latter has the simpler structure it will be considered first. The monomer, ethylene, was originally obtained from sugar by fermentation and dehydration:

$$molasses \rightarrow ethyl\ alcohol \rightarrow ethylene$$

Ethylene is now obtained from oil, or preferably light naphtha, by cracking.

D

6.1.1 *High-density polyethylene*

Two distinct processes were developed almost simultaneously in the mid-1950s, based on Ziegler catalysts and on mixed oxide catalysts: the Phillips process is the most widely employed of the latter.

Ziegler and Phillips HDPEs can be virtually unbranched hydrocarbon chains, leading to densities in the range 0.945–0.965 Mg m^{-3}, although commercial Ziegler polyethylene can have a few ethyl side-groups, (5–7 per 1 000 chain atoms). Oxide-catalysed polymers generally have the highest densities of the polyethylene series.

Processing is dominated by the linearity of the chains, leading to pronounced flow orientation and a rapid crystallization rate with only little supercooling compared with polypropylene. The attainable crystallinity in standard processing operations is somewhat higher than for PP, resulting in generally higher shrinkage. For reasons concerned with processing, HDPE is sometimes preferred to PP for small injection mouldings, where orientation is not a problem, or where it can be tolerated (margarine tubs), and for very large blow-mouldings, where the stability of the parison is helped by the orientation induced during extrusion, and by the pronounced non-Newtonian behaviour of the melt.

The properties of HDPE include a crystalline melting point lower than that of PP, 130–135°C, with very little supercooling and a rapid crystallization rate. HDPE is comparable in rigidity with PP, but with inferior creep behaviour and scratch resistance. Figure 6.1 shows flexural creep modulus as a function of time and the considerable effect of temperature on the deformation behaviour. The effect of density is seen in Figure 6.2, which embraces data covering a density range including LDPE and HDPE. Low-viscosity grades of HDPE are brittle if pronounced stress concentrations are present, but high-molecular-weight materials are tough, even at low temperatures. Low-molecular-weight polymers may stress crack in the presence of some

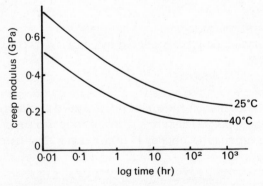

Figure 6.1 Flexural creep modulus as a function of time for HDPE (density 0.953 Mg/m^3) for stress of 4 MPa

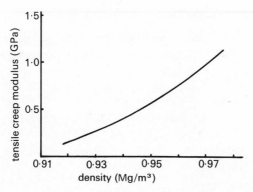

Figure 6.2 Dependence of 100-sec tensile creep modulus at 1% strain on density of polyethylene

chemicals. Density is some 5% greater than PP, with implications for volume cost. HDPE is flammable, although flame retardants improve the polymer in this respect.

As noted above, HDPE is very important for blow-moulding, especially large-capacity containers, such as 220-litre barrels and water butts; dustbins are shaped in this way, two at a time. HDPE is also important for bleach and detergent bottles where the higher rigidity compared with LDPE leads to a cheaper package. Blow-moulding technology is exemplified by a prestigious application of growing importance—fuel tanks for road vehicles. A particularly detailed study has been made of the Peugeot 309 tank (Figure 6.3) which has a mass of 6.5–7.5 kg, a capacity of 48 litres and is produced in a relatively high-molecular-weight HDPE (i.e. the MFI at 2.16 kg at 190°C using a standard die is 0.1 g per 10 min). The requirements, apart from the obvious one of being able to shape the container, include high strength and toughness at −40°C and low permeability to fuel—a loss limit of 20 g per 24 h at 50°C is specified. HDPE, because of its high crystallinity, meets the need for an effective barrier at a mean thickness of 5.5 mm, while the impact requirement specified by the EEC is satisfied by the grades of HDPE developed for this application. As the thickness is 5.5 mm, the moulding cycle is dominated by the cooling time. An increase in productivity can be achieved most readily by reducing the cooling time by providing internal cooling in addition to the normal heat transfer route via the mould. Two aspects are important: increase in the heat transfer coefficient, plastic to air, by circulating the air and provision of sufficient capacity to extract the heat energy. The latter again comprises two factors, the volume of coolant and the temperature at which it is introduced [1].

Reference has been made to injection moulding, which is a fair market for HDPE. Extruded pipe from HDPE and a tougher variant of lower density is widely used for natural gas distribution, particularly in cold climates where some failures of PVC (Chapter 8) have been encountered. HDPE has also been formed into mudguards for commercial vehicles, where the non-corroding

Figure 6.3 Fuel tank for Peugeot 309, blow moulded in HDPE. By permission of RB Blowmoulders Ltd.

polymer withstands the arduous service conditions. Very thin film, made by the tubular process, finds application as a tissue-paper replacement and for point-of-sale wrapping, especially of wet foodstuffs, such as meat and fish. HDPE has been preferred over PP in milk-bottle crates, since creep is less important than for beer crates and the incidence of impact abuse at low temperature is likely to be greater, but the trend is to PP because of its more tolerant moulding characteristics.

HDPE, and more recently, modified polyethylene of lower density, are used in gas distribution systems for many important reasons, including economic shaping, ability to withstand both the internal and external loads, good low-temperature impact properties and relative impermeability to the gases

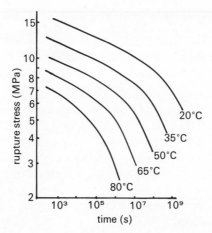

Figure 6.4 Rupture stress vs. time for HDPE pipe at various temperatures

carried. Gas pipes are required to have many years of maintenance-free life, so it is important for the design of the system to be correct. A typical installation requires the pipe to operate for up to 50 years at 20°C at an internal pressure of 0.25 MPa, with a mean diameter of 250 mm. For thin-walled pipes the wall thickness can be calculated, based on a failure stress after 50 years of 6.5 MPa (Figure 6.4). The safety factor suggested in BS 3796 is 1.3; hence the design stress is 6.5/1.3 = 5 MPa and the wall thickness required is 5 mm.

This calculation applies for an ideal pipe, shaped under perfect conditions and tested in a laboratory. However, imperfections may be generated during extrusion or in handling and installation. Gas engineers are anxious to avoid failure caused by a longitudinal crack propagating through the wall and unzipping several kilometres of pipe. A fracture mechanics approach is required, based on a fracture toughness of $3\,MN\,m^{-3/2}$, which has been determined in laboratory tests on this material. The crack will propagate if the rate of energy release resulting from the growth of the crack exceeds the fracture toughness, thus imposing a limiting diameter on the pipe, or a limiting hoop stress (internal pressure), to which the pipe can be subjected. All pipes of diameter less than the critical should be incapable of catastrophic crack growth, or of creep rupture when operated at the specified pressure rating. However, pipes of larger diameter must be operated at a pressure lower than that corresponding to the creep rupture design stress if runaway cracks are to be avoided.

As with polypropylene, reinforcement of HDPE with short glass fibres is practised; compounds with loadings of glass up to 20% are available, giving much improved rigidity and reduced thermal expansion compared with the unreinforced material. The modulus is increased by a factor of 3 for 20% glass fibre but, in the presumed absence of coupling of the glass to the matrix HDPE, the stress at yield is unchanged and may even be reduced in some circumstances.

6.1.2 *Low-density polyethylene*

Low-density polyethylene, generally considered to have a specific gravity in the range 0.915–0.930, is manufactured at high pressure and high temperature, usually 1 500–3 000 bars at 150–250°C, in a process which is usually continuous in stirred reactors or tubes. The reaction conditions invite free radical attack on polymer, leading to long-chain branching and to pendant C_2 and C_4 chains, and to other irregularities, including unsaturation. Molecular weight, molecular-weight distribution and branching are controlled in the reaction by modifiers, by temperature and pressure, and by the type and concentration of initiator.

The structure of LDPE is generally similar to that of HDPE, but the crystallinity is limited by branches which cannot fit into the crystalline lattice. Crystallinity is frequently 50–65%, is also dependent on molecular weight, the higher-molecular-weight grades achieving lower crystallinity. Branching is specified as the number of CH_3 per 1 000 carbon atoms, generally 25–30 for high pressure polymers. Specific gravities of the crystalline and amorphous states are very different, 1.01 and 0.84 respectively, giving a method for assessing crystallinity: X-ray methods are also used. Melt Flow Index (MFI), is still widely used as a monitor of processability: the higher the MFI the greater the fluidity (the lower the viscosity), of the melt. Grades of LDPE suitable for all processing methods, including rotational moulding, are generally available. Notes on the processing methods employed are given in the survey of applications below.

The properties of LDPE are those expected of a lower-crystallinity variant of HDPE; the melting point is lower, approximately 115°C, varying with density, (Figure 6.5), with a more gradual melting process. It is almost a factor

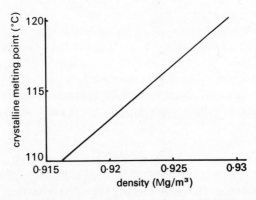

Figure 6.5 Dependence of crystalline melting point of LDPE on density

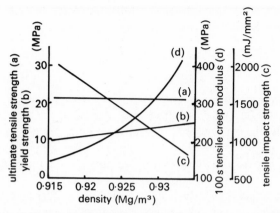

Figure 6.6 Effect of density of LPDE (MFI = 0.25 g/10 min) on some mechanical properties

of 10 less rigid than HDPE or polypropylene and the yield stress is much lower. Both modulus and yield stress depend on density (crystallinity) and on MFI (Figure 6.6, 6.7). At moderate and high molecular weights the polymer is very tough, but low-molecular-weight grades are susceptible to environmental stress cracking. Long-chain branching affects molecular weight vs. processing behaviour favourably. Electrical properties can be excellent, but depend on impurities and additives. Additives are frequently incorporated in LDPE to modify the properties for particular applications; these include flame retardants, slip and anti-blocking additives for film, antioxidants, UV stabilizers, pigments and occasionally prodegradants.

Applications include film for packaging manufactured by the tube-blowing process, which consumes some 70% of LDPE production; a miscellany of uses takes the remaining 30%, including:

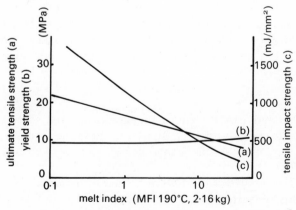

Figure 6.7 Effect of Melt Flow Index of LDPE (density 0.918 Mg/m³) on some mechanical properties

Bottles   Extrusion blow moulding
Mustard jars   Injection blow moulding
Bowls, buckets   Injection moulding
Cold water tanks   Rotational moulding
Paper coating   Extrusion.

For many years, wheelbarrows have been made from wood or metal, both of which have to be protected from the degrading effect of weathering. Unfortunately, the conventional protective coatings are not durable and the substrate is quickly exposed to the elements. The idea of using a plastic moulded bin as part of a wheelbarrow was novel, and allowed changes in the general design. The bin is injection-moulded, so that there is little hindrance to introducing more complex shapes: the design of the barrow was changed so that it was much deeper, with a less acute tipping angle, enabling a larger volume of material to be carried. The feet were connected to the base and the handles placed wide apart to make the wheelbarrow more stable in use. The main performance requirement was that it should withstand the load when full; it should also be durable enough to withstand the impact of objects such as bricks being thrown into it from a short distance. Assuming a maximum load of 150 kg, the approximate stress, assuming a wall thickness of 3 mm, is less than 1 MPa, a stress level which could be tolerated by all thermoplastics: the choice depended, therefore, on impact performance and price.

Low-density polyethylene is still one of the cheapest thermoplastics, possesses excellent impact performance, but is very flexible. A grade of LDPE was selected to give the best overall properties, viz. a density of $0.92\ \mathrm{Mg\,m^{-3}}$ and MFI of 12–14 g per 10 min (2.16 kg at 190°C).

6.1.3 Linear low-density polyethylene

As noted at the beginning of this chapter, the structure and properties of LLDPE are comparable with those of LDPE, with the notable difference that LLDPE does not have long-chain branching. The practical consequence is that LLDPE has much superior strain accommodation, which allows far greater draw-down, both in the melt and in the solid state. It is interesting to compare the tensile stress vs. strain curves for HDPE, LDPE and LLDPE (see Figure 6.8).

The LDPE and LLDPE are of similar density and show very similar behaviour up to yield, with the HDPE having, as expected, a much higher yield stress, fully supporting the thesis that the behaviour to yield is dependent on the crystallinity. However, molecular topography takes over in the later stages of the test. The elongation of LDPE is limited by long-chain branching, and that of HDPE by shortage of amorphous polymer. Not being limited by these two factors, LLDPE can extend to very high elongation.

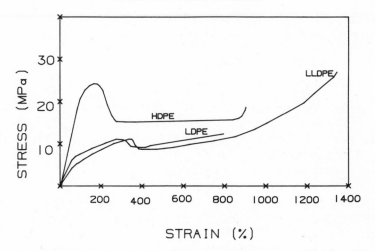

Figure 6.8 Tensile stress vs. strain curves for various polyethylenes

The melt rheology of LLDPE differs from that of LDPE, so that processing equipment for the latter cannot be used without modification. Consequently, much interest in blends has been generated; these are discussed later in this chapter.

Film appears to be the main application area at present.

6.1.4 *Very-low-density polyethylene*

This material, with a density down to $0.88\,\text{Mg}\,\text{m}^{-3}$, is believed to be a copolymer of ethylene with a higher olefin. No pattern of applications seems to have been established.

6.1.5 *Blends of ethylene polymers*

It might seem axiomatic that the various polymers of ethylene discussed above should be miscible; indeed, recently it was recommended that blends of HDPE and LDPE be used to obtain properties intermediate between those of the homopolymers. However, work in several centres has shown that miscibility is the exception rather than the rule: even blends of LDPE and LLDPE tend to become 'unmixed' into separate crystalline phases. This sounds a warning concerning masterbatch carrier polymers; it is not prudent to use a LDPE or, worse, an EVAC copolymer as carrier in a masterbatch introduced into matrices of HDPE and LLDPE, or, worse, polypropylene. It may apparently

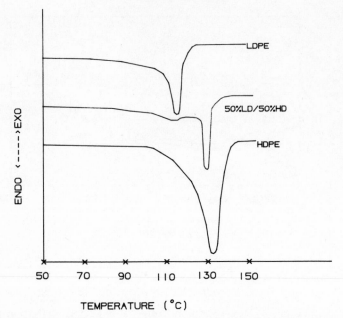

Figure 6.9 Thermograms of LDPE, HDPE and a 50/50 blend of the two

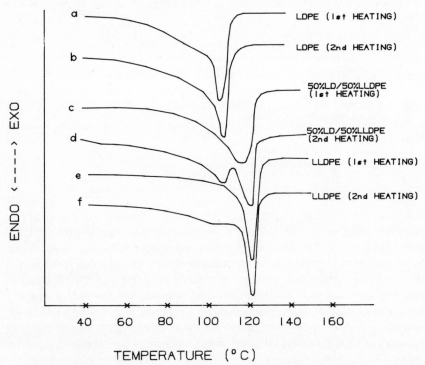

Figure 6.10 Thermograms of LDPE, LLDPE and a 50/50 blend: quenched and slow-cooled specimens

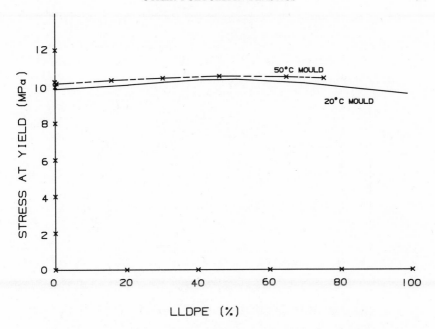

Figure 6.11 Yield stress vs. composition for LDPE-LLDPE blends. (Specimens did not reach yield in the flow direction)

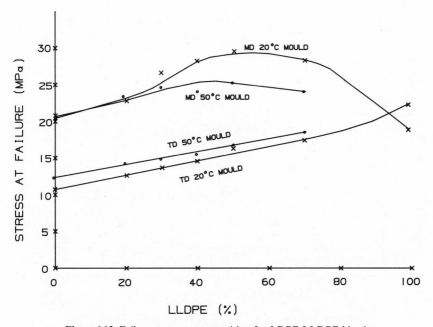

Figure 6.12 Failure stress vs. composition for LDPE-LLDPE blends

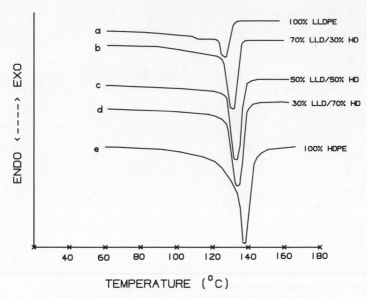

Figure 6.13 Thermograms of LLDPE, HDPE and blends

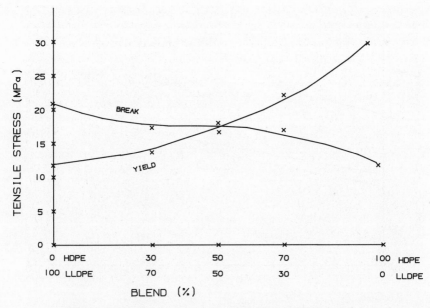

Figure 6.14 Yield and failure stresses for LLDPE, HDPE and blends

be tolerated in a large number of cases, but critical applications may fail. The separation is shown in differential scanning calorimetry thermograms: blends of LDPE–HDPE are shown in Figure 6.9, and of LDPE–LLDPE in Figure 6.10. The difference between Figure 6.10 (*c*) and (*d*) is that (*c*) was quenched from the melt, whereas Figure (*d*) refers to slower cooling. Clearly, for the LDPE–LLDPE system the response is very dependent on the thermal history, and it is no surprise that the mechanical properties are affected.

However, the yield stress in the transverse direction seems to be little affected by the mould temperature (Figure 6.11), with the data being in the order expected, i.e. the quenched samples having the lower yield stresses. A similar picture is presented for the breaking stresses in the transverse direction (Figure 6.12); however, a very different situation obtains in the flow direction, (machine direction), where we find that the considerably higher failure stress is associated with the more rapidly cooled samples. The difference is appreciable, some 15%.

By contrast, the LLDPE–HDPE system appears to be miscible and co-crystalline over the entire composition range; Figure 6.13 shows a steady increase in melting temperature as the HDPE content is increased.

The mechanical properties are generally as expected (Figure 6.14), with the yield stress increasing monotonically with HDPE content, while the breaking

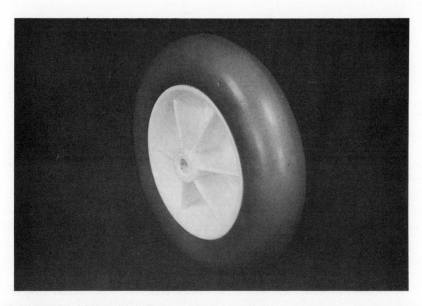

Figure 6.15 Cross-linked microcellular moulded tyre (EVA) on polypropylene wheel for child's push-chair (Andrews-McLaren "Baby Buggy")

stress is sensibly constant, although the value for HDPE seems to be a little lower than the remainder.

6.2 Ethylene copolymers

Ethylene copolymerized with other olefins has already been covered above. Ethylene copolymers with polar comonomers are made by the high-pressure route; the favoured monomer is vinyl acetate (VAC) which, under the usual reaction conditions, copolymerizes with ethylene ideally. Small amounts of VAC reduce crystallinity and melting point, but increase toughness and flexibility: such copolymers (EVAC) are used for tough film, although the moisture permeability is increased by the inclusion of the polar groups.

EVAC was chosen for beer pipes since it is sufficiently flexible, is harmless in direct contact with foodstuffs, and does not taint the beer.

High-molecular-weight copolymers of higher VAC content (20–40%) find a variety of uses as thermoplastic rubbers, particularly as microcellular soling for shoes and as moulded, cross-linked, foamed tyres for the Andrews-McLaren Baby Buggy™ (Figure 6.15). Low-molecular-weight grades are used as reinforcing additives for wax coatings, hot melt adhesives, etc.

Ionomer resins are ethylene copolymers with an active functional group in the comonomer, frequently carboxyl. This is 'neutralized' with metal oxide to give a tough thermoplastic rubber.

6.3 Polybut-1-ene

Up to 1957, only the non-crystallizable (atactic) low-molecular-weight polymer was known; this is still used as a viscosity-stabilizing additive for oils. Isotactic, crystalline polymer is generally similar to PP, but the melting point is some 40°C lower. It is manufactured by a process similar to that for polypropylene.

Polybutene is unusual in that there are two crystalline forms, I and II; the latter is metastable but is *always* formed from the melt during normal processing. The transition from II to I occurs at room temperature over 2–5 days, with increase in density, and consequent change in dimensions. Processing and properties are very similar to those of PP, except that the melting point is 130°C (for Type I crystallinity); similarly, the glass transition temperature is some 20°C lower than PP. Polybutene is slightly less rigid than PP, but has a significantly better creep resistance and seems to be immune to environmental stress cracking.

Applications must be tolerant of the crystal transition and benefit from the creep behaviour. Pipes, especially for high-pressure distribution, appear to be the only important outlet, particularly for cold water and gas, and for hot water installations. They are manufactured in diameters up to 460 mm, principally in North America. Polybutene is also abrasion-resistant and finds application for the transportation of aqueous mineral slurries, e.g. gypsum, by pipeline.

6.4 Poly-4-methylpent-1-ene (P4MP)

This polymer has the repeat unit:

$$\sim CH_2.\underset{\displaystyle \overset{|}{\underset{\displaystyle \overset{|}{\underset{\displaystyle CH_2}{CH_2}}}{CH_2}}}{CH}\sim$$
$$CH_3 \quad CH_3$$

The isotactic form is the only polymer of commercial importance. An unusual feature is that bulk polymer is transparent because the molecular chain is only very slightly birefringent. A second unusual feature is that the crystalline density and the amorphous density are very similar. The transparency is further enhanced by nucleation giving finer texture, and by balancing crystalline and amorphous densities even more closely, by copolymerization. The monomer is obtained by the dimerization of propylene.

The principal properties of poly-4-methylpentene are a melting temperature of 230–235°C and glass transition 40–50°C. There is some tendency to brittleness, although P4MP is tougher than glass, but is very susceptible to UV attack. It is the only bulk transparent polyolefin in commercial production or development. The high melting point forces processing temperatures into the degradation range for the polymer. Melt stabilizers and lubricants are employed; even so, it is difficult to mould orientation-free products, and the presence of residual stresses increases the probability of brittle failure at low imposed stresses.

Applications include a number of small items which require polyolefin inertness, reasonable toughness compared with glass, coupled with either transparency or high softening temperature. Examples are: spirits dispensers, automatic milking equipment, car interior light covers and 'cook-in-the-bag' trays. P4MP is an expensive material, but has a low specific gravity, 0.83, making it cheaper on a volume basis than its main competitor, polycarbonate, which is also more susceptible to chemical attack. P4MP can withstand up to 50 'hot air' sterilization cycles of 1 h at 160°C, as recommended by the British Medical Research Council. However, its brittleness makes it unsuitable for many applications; it has only slightly better impact properties than PMMA. Another drawback is that it is unsuited to external uses, or applications where it is subjected to high-intensity UV radiation.

P4MP is suitable for laboratory ware because it has good chemical resistance, transparency and high-temperature performance. For medical ware these same properties are advantageous; in addition, no toxic products are extracted under normal use. In such applications, additives, such as antioxidants, must be chosen appropriately.

References

1. Birley, A.W. *et al.*, *Plastics and Rubber Processing and Applications* **8**, 337 (1986).

7 Other crystalline thermoplastics

The materials covered in this chapter are the crystalline members of the so-called 'engineering thermoplastics'. They are rigid and robust, but somewhat more expensive than the commodity plastics, PE, PP, PS and PVC. Polyamides and saturated polyesters are used principally in fibres production, but plastics applications are important; indeed, some materials have been developed specifically for plastics uses, notably polyamide 11, polyamide 12 and poly(butylene terephthalate). Polyacetals find application as plastics only.

7.1 Polyamides

7.1.1 *General-purpose polyamides*

The important members of the polyamide family, also known as *nylons*, and outline methods of manufacture, are given below.

(i) *Polyamide 6.6 (PA 6.6)*. Obtained from hexamethylene diamine and adipic acid; stoichiometry in reaction is assured by isolation of 'nylon salt'. Melting temperature of PA 6.6 is 265°C.

(ii) *Polyamide 6 (PA 6)*. Manufactured by the ring-opening polymerization of caprolactam: still contains 6% monomer at equilibrium, which is removed by extraction. Melting temperature of PA 6 is 225°C.

(iii) *Polyamide 6.10 (PA 6.10)*. Results from condensation of hexamethylene diamine with sebacic acid: melting temperature 222°C.

(iv) *Polyamide 11 (PA 11)*. Self-condensation of ω-aminoundecanoic acid (obtained from castor oil). Note that cyclic structure is not formed: melting temperature 185°C.

(v) *Polyamide 12 (PA 12)*. Prepared by the self-condensation of ω-amino acid, which is obtained from butadiene trimer. Melting temperature 175°C.

Other polyamides have been developed, including polyamide 4.6 (Dutch State Mines) which, as expected, has a melting temperature higher than that of PA 6.6, and is intended for service at temperatures above the ceiling for PA 6.6.

Polyamide MXD-6 is the product of condensing metaxylylene diamine with adipic acid. This polymer has proved to be an outstanding gas barrier; for example, at 100% relative humidity, it has lower oxygen permeation than ethylene–vinyl alcohol copolymer (30% ethylene). Otherwise, it is similar to poly(ethylene terephthalate), with melting temperature 243°C and glass transition temperature 64°C, and very similar crystallization behaviour. However, the polymer does absorb moisture at a level comparable with PA 6.6. PA MXD-6 is finding application is stretch-blow moulded bottles, sandwiched between poly(ethylene terephthalate). Aromatic polyamides and polyimides will be covered later in this chapter.

Proteins are based on amino-acid condensates, i.e. they are polyamides, but they differ from the simple polyamides listed above, consisting of many 'monomers' in a sequence which is unique to a particular protein.

Polyamides of higher solubility can be made by copolymerization, which restricts crystallization. The common polyamides PA 6.6 and PA 6 are frequently of low molecular weight; the resulting low viscosity requires special precautions during processing. Extrusion is important for PA 11 and PA 12. Polymers must be scrupulously dry before processing, otherwise hydrolytic degradation occurs, with consequent deterioration in properties.

(i) *Properties.* The properties of crystalline polyamide plastics differ from those of the polyolefins, considered in the previous two chapters, consequent on the inclusion of the highly polar *amide* linkage in their structure.

$$\sim C - N \sim$$

PA 6 and PA 6.6 have high crystalline melting temperatures as noted above, and all polyamides are water-attractive, the water absorption (and the melting temperature) decreasing as the methylene: amide ratio increases, ranging from 8–10% for PA 6 and PA 6.6 to 2–3% for PA 11 and PA 12 at saturation. The absorption of water affects dimensional stability, each 1% of water absorbed resulting in an increase in linear dimensions of 0.3% although this may be partially compensated by post-moulding shrinkage (Figure 7.1). The deformation behaviour is considerably affected by moisture, the effective 'modulus'

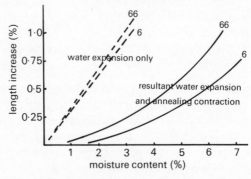

Figure 7.1 Polyamide 6 and 6.6: change in dimensions with water content

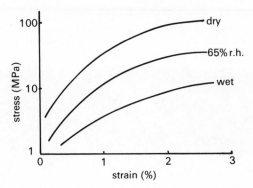

Figure 7.2 Polyamide 6.6: effect of water content on isochronous stress vs. strain properties

being reduced to 20–25% of its dry value for equilibrium 'wet' material (Figure 7.2).

Polyamides have good strength and toughness and excellent fatigue resistance, the toughness being increased markedly by absorbed water; indeed, PA 6.6 is brittle when thoroughly dry (Figure 7.3). Electrical properties are not outstanding, even for dry polymer, and deteriorate further on absorbing moisture. In chemical properties, polyamides are attacked by acids, but are stable to alkalis; they are resistant to hydrocarbons, esters and glycols, but are dissolved by strongly hydrogen-bonding solvents, e.g. phenol.

(ii) *Applications.* PA 6 and PA 6.6 are used mainly in textiles but find many plastics uses, usually where toughness is a prerequisite: some examples are oil-filler caps for road vehicles; teeth in plastics zip fasteners; castors for light furniture; hopper barrels for food mincers; radiator tanks for cars; hose connector for Electrolux vacuum cleaner; and gears, especially in food processing equipment.

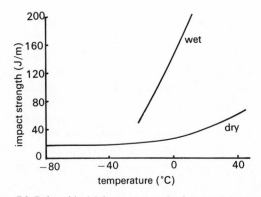

Figure 7.3 Polyamide 6.6: impact strength of wet and dry samples

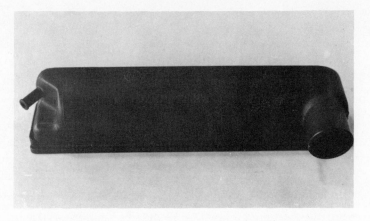

Figure 7.4 Polyamide 6.6: heat-stabilized grade in car radiator header tank

In spite of indifferent electrical properties, PA 6.6 is used by British Rail to insulate rails from concrete sleepers, as part of the track circuitry for signalling. It finds wide application in switch components in telephone exchanges, and in connector harnesses in car wiring. A radiator header tank (Figure 7.4) uses PA 6.6, chosen in respect of its high heat distortion temperature, since the component frequently experiences temperatures above 100°C and pressures of 90 kNm^{-2}.

In spite of a price significantly higher than that of the 'commodity' polyamides, PA 11 and PA 12 offer a combination of properties which allow their economic use in a variety of applications. PA 11, which is the longer established of the two, is more flexible than PA 6.6, is tougher, and is less affected by moisture. Applications include petrol pipes for cars, hydraulic pipe, brake fluid reservoirs and film for packaging cooked meat.

7.1.2 Modified polyamides

(i) *Fibre-reinforced polyamide.* While the cheapest of the polyamides, PA 6 and PA 6.6, offer attractive properties at a competitive price, they have shortcomings which have been the targets of considerable applied research, the results of which can be seen in a number of modified grades. The principal deficiencies are associated with water absorption, and include dimensional instability, low rigidity at environmental humidity, and a tendency to brittleness at low humidity. The first two of these were improved by the introduction of short (2 mm) glass fibres, grades containing 15–40% filler giving improved solid-state properties, but retaining the capability of being processed as thermoplastics. The processing of such glass-filled grades is generally similar to that of the unfilled polymers, but mould design must be considered very carefully to accommodate anisotropy resulting from flow orientation of the glass fibres. Mould shrinkage is much reduced, but anisotropy in shrinkage is usually introduced.

The properties of the glass-filled grades depend on the glass fibre content; the comments below, and many of the applications cited are for PA 6.6 containing 33% of glass fibres. The specific gravity is 1.39, compared with 1.14 for the unfilled grade and the saturation water content is reduced from 8.5% to 5.8%. The rigidity is increased by a factor of three, even at strains of 2–3%, and the strength is enhanced by a similar factor.

Applications are frequently characterized by concurrent requirements of high rigidity to locate moving parts precisely, and of easy moulding to complex shapes. Electrical insulation and toughness are added advantages. Examples of successful applications include: chain-saw housings; body of the Kango™ electric hammer; cooling fans for commercial vehicles (where peripheral speeds up to $100\,\text{m\,s}^{-1}$ can be tolerated); and a casing for a portable electric drill, which is discussed in greater detail below. Perhaps less exacting applications include handles for garden secateurs, electric cooker control knobs, and a concert clarinet.

The anisotropy resulting from the incorporation of glass fibres in PA 6.6 may be less acceptable to the designer than the somewhat smaller improvement in mechanical properties obtained by addition of up to 40% particulate mineral filler. The particulate filler gives considerably lower strength and rigidity compared to glass-fibre-reinforced polyamide, but mould shrinkage and the effect of moisture on dimensional stability are comparable, and the processing difficulties associated with the glass fibre filled grade are considerably reduced. However, as often found for particle-filled compounds, the brittleness is increased significantly.

Early designs of hand-drills involved manufacture of the casing from several parts. The body was cast, but problems with restricted flow in the casting operation limited the complexity of the component. The advantage offered by glass-fibre-filled polyamide was that it could be moulded into a very complex shape in a single operation, without the need for further machining. Furthermore, plastics provided a safer product and a reduction in weight. The main requirements in this product were high strength and rigidity, with good creep resistance, especially at elevated temperatures, good toughness, close moulding tolerances, with good dimensional stability, and electrical insulation. An example is shown in Figure 7.5.

Unfilled polymers could be discounted immediately as having insufficient strength and rigidity and limited creep resistance at high temperature. Generally, non-fibrous fillers increase rigidity but not strength, so only fibre-reinforced polymers are suitable. Of eligible materials, ABS, PS-modified PPO and PP have low strength and heat deflection temperatures, although they possess good impact properties; filled acetal has properties inferior to polyamide, polyethersulphone and thermoplastic polyester, and is not considered further. Polycarbonate is tough but has too low a heat deflection

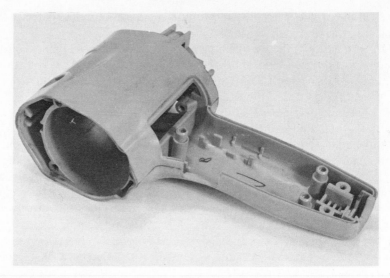

Figure 7.5 Electric drill housing in glass fibre-reinforced polyamide 6.6 (Black and Decker)

temperature for this application. Glass-reinforced polyamide was chosen as the most suitable material, since it is less costly than polyethersulphone and gives a better surface finish than glass-reinforced thermoplastic polyester. Cross-linked plastics generally have unsuitable impact characteristics and insufficient melt flow to produce this complex component. The high heat-deflection temperature is necessary because the brush assembly for the electric motor is connected directly to the plastic drill housing and may be subjected to a temperature above $200°C$. Polyamide has the problem that it is hygroscopic, but with the addition of glass, the maximum amount of water absorbed is 5.8%, with a corresponding dimensional change not greater than 1.5%. In a normal room environment at $20°C$ and 65% relative humidity, the equilibrium moisture content will be 2% for PA 6.6 with 33% glass fibres. In applications where closer tolerances are required, glass-reinforced thermoplastic polyester or polyethersulphone would have to be considered.

(ii) *Elastomer-modified polyamides.*   The tendency of PA 6.6 to brittleness at low humidity has been a hindrance to its use in some engineering applications. Grades are now available in which this has been overcome by the addition of rubber, (frequently ethylene-propylene copolymer elastomer) inevitably with some decrease in rigidity and strength. The rubber-modified product already has an impressive list of successful applications, including a vacuum-cleaner impeller, motor-cycle drive sprocket, handle for rotary hammer and gear for the starter mechanism of a lawn mower. The same toughening technique appears to be possible with mineral-filled PA 6.6, the considerable improve-

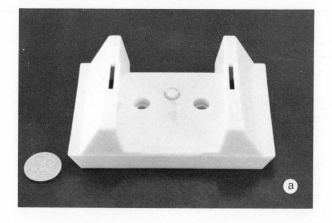

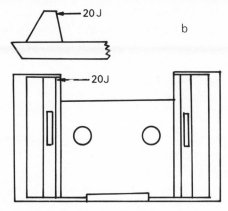

Figure 7.6 (*a*) Sealer bed pocket in polyamide 6.6 ST (super tough) moulded by Molins Ltd., London (*b*) Application of impact force

ment in impact behaviour allowing use of the compound in automotive parts. Toughening is again accompanied by some loss in rigidity and strength.

The sealer bed pocket (Figure 7.6.*a*) is an important component in a cigarette-packaging machine manufactured by Molins Ltd., London. It collects into bundles cigarettes which have already been rolled, wraps them in foil, and forms a cardboard pack around the bundle, operating at a speed of 200 packs per minute. The sealer bed pocket is used to locate the cardboard while the card is wrapped around the bundle of cigarettes. Several parts of these machines are made from a special grade of aluminium, which can be machined easily; however, many of these are not subjected to high stress. The sealer bed pocket was an application for which plastics seemed to offer an improvement, since machining the part from aluminium was time-consuming and produced large quantities of expensive scrap.

When the packaging unit is in use, a pack of cigarettes occasionally becomes

Table 7.1 Impact properties of polyamide materials

Pocket material	Impact energy at failure (J)
PA 11	2.6
PA 11 (glass sphere-filled)	1.7
ABS	8.0
Impact modified PA 6.6	21.3 (unbroken)

lodged between the pocket and the card-folding mechanism, causing the aluminium pocket to break. The impact load was estimated to be 20 J, occurring at the extremity of the two protruding ears of the pocket (Figure 7.6b). The pocket material requires good wear properties, as the cardboard slides over the surface at very high speed; further, there is the possibility of loose tobacco and other cigarette additives becoming sandwiched between the card and the pocket, causing abrasion. Finally, the pocket must fit well between the guides of the machine to prevent twisting. It was considered that the most important criterion in materials choice for this application is the ability to withstand the impact loading. Impact tests on four materials, simulating the stressing mode, gave the results in Table 7.1, resulting in the selection of a high impact polyamide 6.6.

Although the impact-modified PA 6.6 gave the best impact performance, this is slightly offset by the greater distortion in mouldings and the decrease in strength and rigidity, consequent on toughening (Table 7.2). The component has been designed with constant section thickness of 3 mm, to reduce distortion and improve moulded dimensional tolerances. The material is fed through a sprue, which is recessed away from the main sliding surface, to enable the sprue to be drilled out easily without interfering with the main sliding surface.

Monomeric and oligomeric forms of caprolactam modified with a compatible elastomer (such as butadiene) and usually glass-fibre-reinforced, can be conveniently processed using reaction injection moulding equipment to produce block copolymers of PA 6, (e.g. Dutch State Mines' Nyrim™). Applications include panels for the underside of snowmobiles.

Table 7.2 Properties of dry polyamides

Property	PA 6	PA 6.6	PA 11	Impact-modified PA 6.6
Tensile strength (MPa)	86	86	58	52
Elongation (%)	20 to 30	40 to 80	300 to 350	–
Flexural modulus (GPa)	2.75	2.85	1.00	1.76
Notched Izod impact (J/m)	53	70	210	800
Mould shrinkage (%)	0.6 to 1.8	1.3 to 2.3	–	1.5
Saturation water absorption (%)	9.0	8.5	1.8	6.7

7.1.3 *Aromatic polyamides and polyimides*

Wholly aromatic polyamides derived from aromatic diamines, such as phenylene diamine, and aromatic dicarboxylic acids, such as one of the phthalic acids, are extremely high softening polymers, which are not necessarily crystalline: they find application as speciality fibres. Nomex™, derived from *m*-phenylene diamine and isophthalic acid, is used for re-entry parachutes for space vehicles and in textiles for firefighters' apparel. Kevlar™, another condensate of aromatic diamine and dicarboxylic acid, is one of the stiffest and strongest organic fibres. Since this polyamide contains *para*-linked rings exclusively, it is symmetrical and consequently crystalline. Its processing presents special difficulties, which have been overcome in the case of the fibre by spinning it in the liquid crystal state, the diluent (or 'solvent') being sulphuric acid. It is difficult to visualize how this technology could be extended to plastics use, although Aramid™ resins are available as stock shapes.

Polyimides, again based on aromatic precursors, are noted for their high-temperature performance, and find some use as high-temperature electrical insulants. Polyimides are, in one sense, similar to cross-linked plastics in that they are processed as prepolymers; the final stage of the reaction yields a product which is still strictly a linear polymer but which degrades before softening at temperatures over 500°C. This structure allows continuous use in the absence of air at temperatures up to 315°C with transient surges up to 480°C. However, unlike most cross-linked plastics, this polymer is claimed to be tough; it is available in semi-fabricated forms from which end-users can machine their own parts.

7.2 Thermoplastic polyesters

7.2.1 *Homo-polyesters*

The two most common thermoplastic polyesters are poly(ethylene terephthalate) (PETP), melting point 265°C: and poly(butylene terephthalate) (PBTP), melting point 230°C. These two polymers are accessible because of the widespread use of PETP in fibres and films: they are produced from ethylene glycol or butylene glycol respectively, condensed with terephthalic acid or dimethyl terephthalate. The attributes of these polyesters are properties comparable with PA 6 and PA 6.6, but with considerably lower water absorption, with beneficial effects on the dimensional stability of mouldings, and in the retention of mechanical properties in short-term contact with moisture. PETP presents processing problems in that crystalline products are possible only at mould temperatures of 140°C, due to quenching at lower mould temperatures, although a variant overcoming this deficiency is now claimed. PBTP, also known as polytetramethylene terephthalate (PTMT), is not so restricted, and crystalline products are obtained directly at

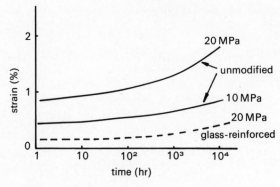

Figure 7.7 Poly(butylene terephthalate): tensile creep at 20°C

conventional mould temperatures. A disadvantage of thermoplastic polyesters is their indifferent hydrolysis resistance: prolonged contact with water at 95°C, or even at 50°C, has a significant detrimental effect on properties. These polymers are characterized by their dimensional stability, being largely unaffected by changes in the ambient humidity, and by high rigidity, which is retained to high temperatures. This is particularly true of the glass-fibre-reinforced grades: illustrative data are given in Figures 7.7 and 7.8.

Although PETP is not widely used as a plastics material, two applications, one long established and one a more recent development, are important. These are biaxially oriented film and biaxially oriented bottles, which are suitable for carbonated drinks. In the manufacture of both products, the following stages apply:

 (i) Melt and homogenize
 (ii) Primary shaping (film extrusion or moulding a stubby parison)
(iii) Quench to substantially amorphous state
(iv) Reheat and orient biaxially, sequentially or simultaneously
 (v) Maintaining orientation, heat to crystallize; sets orientation.

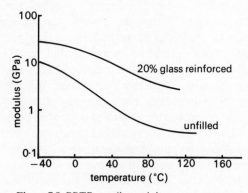

Figure 7.8 PBTP: tensile modulus vs. temperature

The purpose of the orientation is to improve the barrier properties, while allowing thinner bottles to be made; rigidity, strength and toughness are also enhanced.

Poly(butylene terephthalate) (PBTP), a much more recent introduction than PETP, has already found a number of outlets, including the bodies and stems of push-button switches on Jaguar cars, valve components for automatic cattle waterers, and motor mountings in refrigerators for beer chillers. By far the most important grades are the glass-reinforced plastics, which probably account for 95% of the PBTP used in mouldings. Examples include terminal boxes for electricity supply, sextant components, centrifugal pump components, parts for a sandwich grill, base and stand for coffee-making machines and motor end plates in the Vent-Axia 150 domestic extractor fan. Two of these are considered in greater detail to illustrate the versatility of glass fibre-filled thermoplastic polyesters; the Vent-Axia fan and the Rima-contact grill. It appears that grades of PETP, specially formulated for more rapid crystallization, offer enhanced stiffness and strength, compared with PBTP, suggesting use in a number of applications which are currently held by PBTP.

In 1977, the Vent-Axia 150 was developed for the domestic market: special attention was given in the design to enable manufacture by the cost-effective injection moulding of most parts. At the design stage, the parts were separated into two categories: the aesthetic envelope parts, which are subject to minimal stresses (Figure 7.9), and the mechanical parts, requiring mechanical rigidity and strength (Figure 7.10). The most important mechanical part is the centre moulding, because it supports the weight of the electric motor and shutter, as well as providing accurate location for the rotor and stator of the motor. Thermosets were discounted for this component, since they are difficult to

Figure 7.9 Vent-Axia fan: aesthetic envelope parts moulded by Derwent Plastics Ltd

Figure 7.10 Vent-Axia fan: mechanical parts moulded by Derwent Plastics Ltd. Centre moulding is arrowed

mould with intricate detail: thermoplastics materials most suited are fibre-reinforced versions of polyamide, polycarbonate, thermoplastic polyester, polyethersulphone, polypropylene, modified polyphenylene oxide and ABS. Comparison of the mechanical properties of these materials reveals that polyamide and thermoplastic polyester provide the best balance of properties, including cost. The centre moulding is always under stress, so creep is another important factor (Figure 7.11). Other factors affecting choice were dimensional stability and post-moulding dimensional consistency. Glass fibre-

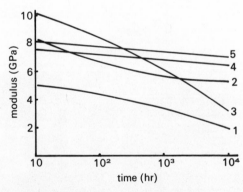

Figure 7.11 Tensile creep moduli of glass-reinforced plastics

1. 30% glass-reinforced polypropylene $32°C: \sigma = 11.3\,MPa$
2. 30% glass-reinforced polycarbonate $23°C: \sigma = 13.8\,MPa$
3. 30% glass-reinforced polyamide 6.6 $20°C: \sigma = 20\,MPa$
4. 30% glass-reinforced acetal $20°C: \sigma = 10\,MPa$
5. 30% glass-reinforced polyester (PBTP) $20°C: \sigma = 20\,MPa$

reinforced thermoplastic polyester was chosen because it has excellent mechanical and electrical properties and good fire retardancy.

On the rear side of the centre moulding, a gear needs to be accurately located, and must rotate easily in the recess provided. This gear meshes with gear teeth on the seven shutters, each of which is located on one of the seven studs protruding from the side of the moulding. Thus the tolerances and clearances between these parts must be controlled accurately to allow the shutter mechanism to operate satisfactorily. The limits and fits are specified at the design stage but it is important to appreciate the effects of shrinkage and after-moulding tolerances. Further, the front side of the moulding has critical dimensions because it locates both the rotor and stator of the electric motor, and efficient operation depends on a small and consistent gap between these components. For glass-reinforced polyester, shrinkage depends on the wall thickness of 2.5 mm; the shrinkage in the flow direction is 0.25–0.35%, and in the transverse direction 0.5–0.7%. Tolerances of $\pm 0.2\%$ are possible providing the moulding conditions are accurately controlled, and with extreme care, consistency may be improved to $\pm 0.1\%$, particularly for parts less than 25 mm in their largest dimension.

A different selection of properties is required in a number of components for a contact grill manufactured by Rima Ltd, giving considerable savings in cost and weight, compared with traditional materials (Figure 7.12). Glass-filled PBTP mouldings provide handles, legs and end-plate unit, hinge bearings, front handles, temperature control escutcheon and the temperature selector lever and knob for this electric grill. Material selection required high temperature resistance and retention of high rigidity to high temperatures; good impact strength, low thermal conductivity and good chemical resistance

Figure 7.12 Glass fibre-reinforced PBTP components in contact grill (Rima Ltd)

to fats, oils and other domestic chemicals. Intricate design of the mouldings is possible with the easy-flow plastics materials, and hot runner systems reduce wastage. The largest moulding comprises a carrying handle, leg supports and an end-plate in a single unit, replacing four separate metal components in the original design.

Other thermoplastic polyesters have been reported, including copolyesters based on 1, 4-cyclohexylenedimethylene glycol and terephthalic acid; and polymers and copolymers of p-hydroxybenzoic acid. Little is known about established applications. Hot melt adhesives, based on polyesters made from adipic or phthalic acids with mono- or di-ethylene glycols or 1, 4-butane diol, find use in shoe manufacture.

7.2.2 Blends of polyesters

Blends of PBTP and polycarbonate (PC), marketed by General Electric (USA) under the name Xenoy™, overcome some of the deficiencies of PC, in particular, improving the processing behaviour, hydrolysis resistance and low-temperature impact performance. The amorphous regions of the PBTP appear to be miscible with the wholly amorphous PC, with the glass transition of the 50:50 blend at 100°C being intermediate between that of PC (150°C), and that of PBTP (35–40°C). The PBTP crystallizes to a small extent. The impact behaviour is improved by the inclusion of a small quantity of acrylic rubber. A demanding application for the 50:50 blend, with added rubber, is the

Figure 7.13 Front end of Ford "Sierra" in Xenoy CL 101™

front and the rear end (fenders) of the European Ford Sierra, illustrated in Figure 7.13.

The blend, when properly processed, offers toughness down to $- 70°C$, is tolerant of recycling, (provided scrupulous care is taken over drying), and is reasonably resistant to hydrolysis, as shown by immersion in deionized water at 40°C for 30 weeks.

Finally, blends at different blend ratios are available, e.g. 80:20 PBTP:PC, which is recommended for improved high-temperature performance (see also Chapter 11).

7.3 Polyacetals

There are two variants; the *homopolymer*, which is end-capped with acetate groups and melts at 175°C, and the *copolymer*, containing $-CH_2-CH_2-O-$, melting at 163°C.

The homopolymer is polymerized from formaldehyde, and the copolymer is obtained by copolymerization of, for example, trioxane (a trimer of formaldehyde) with ethylene oxide, or dioxane. The homopolymer is stabilized by end-capping by esterification (treatment with acetic anhydride), or etherification (dimethyl sulphate). The copolymer is degraded by treatment with aqueous ammonia to the stable $-CH_2-CH_2-OH$ end group, which is resistant to further unzipping. A consequence of these stabilization methods is that the homopolymer continues to degrade if the chain is broken, or an end-cap removed, whilst the copolymer degrades only to the next comonomer residue; also the copolymer is considerably more stable to hydrolysis. On the other hand, copolymerization restricts crystallization, giving the copolymer lower rigidity and strength than the homopolymer, and the lower melting temperature recorded above.

In processing, the most important consideration is that overheating leads to the prolific liberation of formaldehyde, which is unpleasant and can be extremely dangerous. Although polyacetals are less hygroscopic than polyamides, the materials must be dry for processing. Crystallization from the melt is extremely rapid, with negligible supercooling. Low melt strength leads to problems in blow-moulding; indeed, polyacetals are mainly used for injection-moulding and injection blow-moulding. Polyacetals (POM from polyoxymethylene), have high stiffness and strength, are reasonably tough and have good dynamic fatigue resistance (Figures 7.14, 7.15). The homopolymer is some 10–15% more rigid and stronger than the copolymer and is comparable in impact properties.

The almost simultaneous development of the homopolymer and copolymer has resulted in competition for the available markets, with more competition with the advent of the thermoplastic polyesters. Generally, the greater strength and rigidity of the homopolymer promote its preferential use in cams and gears (illustrated by the case study below); window-winding handles for cars,

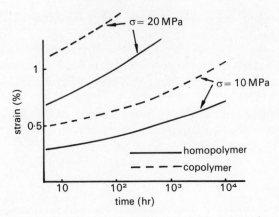

Figure 7.14 Tensile creep of acetal homopolymer and copolymer

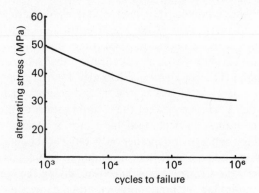

Figure 7.15 Fatigue resistance of acetal homopolymer

and exterior door-handles for cars and trucks. The copolymer, on the other hand, is often preferred where chemical resistance and especially hydrolysis resistance is important: examples include wash basins, taps and other plumbing items; electric kettles; and the mincer attachment for an electric food mixer.

The screen-wiper assembly comprises three units: electric motor, gearbox and oscillating wiper mechanism. We are concerned here with one element of the gearbox, the wormwheel (Figure 7.16). The system operates initially from the fast rotation of the electric motor, which is reduced by a worm gear on the motor shaft driving a plastic wormwheel gear. The main mechanical requirements for the gearwheel were:

Continuous torque	8 J
Maximum excursion temperature	140°C
Maximum continuous running temperature	93°C
Gear ratio (approximately)	50;1
Loading period, continuous	500 h

Figure 7.16 Windscreen wiper worm gearwheel in acetal homopolymer moulded by Lucas Electrical Ltd. Left, an early design; right, modified tooth design for higher torque

The gearwheels were designed to fulfil the 50:1 ratio, and to be as compact as possible. The stresses at the base of the gear teeth can be estimated using beam bending theory: for the teeth in Figure 7.16,

Width of tooth (w)	= 6.35 mm
Addendum and dedendum (l)	= 1.98 mm
Thickness of tooth at base (d)	= 2.61 mm

Force on each gear tooth (W) $= \dfrac{\text{Stall torque}}{2 \times \text{radius of gear wheel}}$

$= 148.2\,\text{N}.$

The stress at the base of the gear teeth can be calculated from:

$$\frac{\sigma}{y} = \frac{M}{I}$$

where

M = bending moment $\qquad\qquad\qquad\qquad = WI$

I = second moment of area $\qquad\qquad\qquad = \dfrac{bd^3}{12}$

Y = maximum distance to top or bottom of beam

$= d/2$ (for rectangular beams only) whence $\sigma \qquad = \dfrac{6WI}{bd^2}$

$= 40\,\text{MPa}.$

The gearwheel must withstand this load after 500 h of continuous operation;

E

the materials which will survive this loading, excluding glass-filled materials which will wear badly, are PC, PA, POM, thermoplastic polyesters, and polysulphones. Cross-linked plastics are too brittle to be considered for gears; polycarbonate and polysulphone have poor fatigue resistance; polyesters are suitable, but, being more expensive than polyamides or polyacetals, they are used less frequently.

Polyamide has superior mechanical properties compared with acetal, but has the serious limitation that it is hygroscopic. When processing polyamide the water content must be extremely low; thus, after moulding it contains very little water and is rigid and brittle. However, PA 6.6 will reach equilibrium with its environment; in normal conditions of 23°C and 60% relative humidity, the water content will be about 3%. The ingress of water reduces the rigidity and strength, improves the impact performance but most important, the dimensional changes are such that it is unsuitable for gears or for any component requiring close tolerances when operating in various humidity environments. Acetal is much more stable dimensionally and possesses excellent wear and fatigue resistance; thus it is often specified for gears. Thermoplastic polyesters are used when the application demands better performance which can justify the increased cost.

Experience in the design of wormwheels was invaluable during this

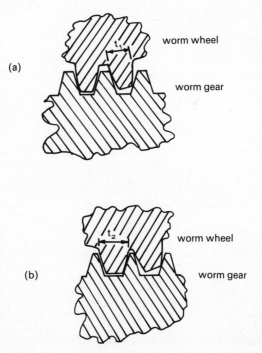

Figure 7.17 Acetal wormwheel; (a) an early design; (b) modified tooth design for higher torque

Table 7.3 Properties of materials for plastic kettle body

Material	Heat deflection temperature (°C) @ 0.46 MPa	@ 1.81 MPa	Maximum water absorption (%)	Relative cost
Acetal copolymer	158	110	1.6	1.0
Polyamide 6.6	190	75	8.5	1.5
Polycarbonate	145	135	0.6	1.5
Polyethersulphone	—	203	2.1	5.5
PS-modified PPO	148	103	0.3	1.5
Thermoplastic polyester	179	59	0.7	1.5

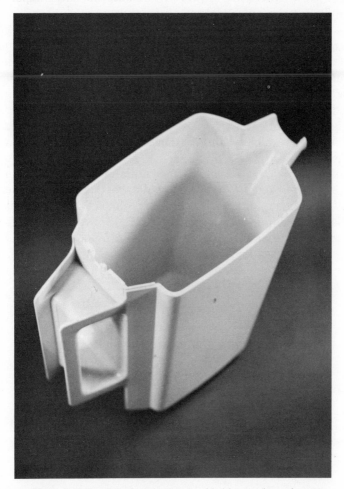

Figure 7.18 Kettle body, injection-moulded in acetal copolymer

development, but can be stultifying when changing from metal to plastic. The first system was a near-identical plastic replacement for metal; a better design could have been produced if the load-carrying capacity of the parts had been balanced. Instead, a gearbox was produced which contained a plastic component, stressed to its limit, and a steel worm, which was lightly stressed. In later designs requiring more torque throughput, the gearbox was redesigned: the steel worm was reduced to its minimum size, allowing larger and thicker wormwheel teeth (Figure 7.17). This allowed the gearbox to transmit 50% more torque.

Amongst a miscellany of applications of polyacetals may be cited tenor recorders: zip fasteners, (replacing metal or polyamide); roller chain and other conveyor components; counter wheels on petrol pumps; and business machines. Polyacetals have excellent fatigue resistance, illustrated by their use in automobile direction-indicating switches.

Metal-bodied kettles are the result of many manufacturing operations, and the advantage of a plastic is that it can be moulded in one piece to provide features to which to attach the electrical element and socket, and provide a colourful finish, at a cost considerably less than that of the metal part. Obviously the main requirement is to withstand boiling water; on average a kettle is boiled for three-minute periods ten times every day. Furthermore, it should be resistant to attack by fats, oils and other kitchen chemicals, have good dimensional stability, and resistance to thermal cycling and to stress cracking.

The most suitable materials on temperature considerations are listed in Table 7.3. Cross-linked plastics would be suitable thermally and chemically, but their flow limitations and brittleness prohibit their use. Of the three materials suitable for use in boiling water, polyethersulphone is very expensive, and PPO is attacked by some chemicals found in kitchens. Acetal is, therefore, the most suitable (Figure 7.18); and although the heat stability of the material is suspect for continuous operation at 100°C, (Figure 7.19), it is

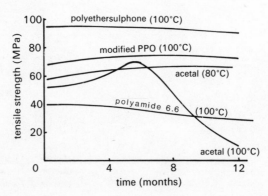

Figure 7.19 Effect of storage in water at high temperature on tensile strength of various plastics

relatively unaffected by short excursions to this temperature. Acetal homo-polymer is not suitable for contact with water above 60°C; however, the copolymer is satisfactory and has been used since 1969 for hot-water plumbing applications. An advantage over metal is less scaling, and grades are available which satisfy the toxicity requirements of the United States Food and Drugs Administration. Another candidate material for this application is impact-modified polypropylene.

Glass fibre-filled grades are available with enhanced rigidity and strength, but they do not appear to be used widely.

8 Vinyl chloride plastics

Poly(vinyl chloride), PVC, is one of the most versatile of polymers, and the following forms have important applications: rigid (unplasticized) grades, plasticized compounds, copolymers and blends. This is perhaps surprising since it is one of the least stable polymers, for which reason much of the early development was concerned with copolymers and with plasticized compositions.

The monomer is now almost exclusively obtained from ethylene, which has displaced the older acetylene-based process. There is some evidence that vinyl chloride is carcinogenic and there are legal requirements limiting exposure to the monomer.

Polymerization is by free-radical-initiated reaction, in either suspension or emulsion processes in water at about 50°C. In the suspension process, polyvinyl alcohol, incompletely hydrolysed PVAC, or gelatine, are used as a colloid stabilizer with peroxide initiators. The emulsion process employs emulsifiers, such as sodium lauryl sulphate and water-soluble persulphate initiators. Mass polymerization gives PVC powders with characteristics different from those of suspension polymer.

Whilst the chain linkage is substantially head-to-tail, there is still controversy over the amount of long-chain branching involved. Commercial PVC has a small amount of crystallinity (about 5–15% depending on the thermal treatment), being syndiotactic in nature. It has an important influence on properties, not always beneficial, and must be considered explicitly in processing and implicitly in data for design. The crystallizability can be increased by polymerizing at low temperature, but such polymers are not commercially important. Molecular weight is an important variable, being recorded as the K *value*, which is derived from solution viscosity measurements. K values from 45 to 90 (cyclohexanone) cover number-average molecular weights from 25 000 to 85 000.

8.1 Unplasticized poly(vinyl chloride) (UPVC)

Because of its limited thermal stability together with the low amount of crystallinity, the melt viscosity of PVC is very high at normal processing temperatures. Therefore, thermal stabilizers and lubricants are essential ingredients for processing. Instability is usually assessed by colour change, resulting from hydrogen chloride evolution which leaves an unsaturated chain, although degradation is a complex reaction in which oxygen is involved.

Some stabilizers are effective in improving weathering behaviour. Lead compounds are good heat stabilizers; basic lead carbonate or tribasic lead sulphate are favoured stabilizers but are toxic. The stearate, palmitate and octoate salts of Cd, Ba, Ca, Zn are increasingly used, but toxicity limits the range of applications for Cd. Organo-tin compounds are also effective stabilizers, especially for transparent products, but may also be somewhat toxic.

Successful processing of unplasticized or rigid PVC (UPVC) requires lubricants, such as lead or calcium stearates, stearic acid or hydrocarbon waxes. UPVC is frequently extruded, and so lubricants melting at 100–120°C are most effective for extrusion at 165°C. UPVC is also widely used in sheet form, obtained by compression moulding multiple plies of calendered foil.

Sheet is manipulated by heating and shaping (thermoforming); it is frequently welded or cemented with solvent-based adhesives. UPVC compound can be injection-moulded using a screw pre-plasticizer.

PVC is a rigid material and is tough under ambient conditions if stress concentrations are absent (see Figures 8.1, 8.2). The softening point is 85°C, but PVC is not reliably load-bearing above 60°C. It is resistant to water and hydrocarbons, but is soluble in polar solvents, although high temperature may be required to effect solution. Typical applications include pipe, rainwater goods (by extrusion) with fittings (by injection-moulding); sheet used in hygienic cladding; extruded sheet sections which lock together, e.g. to form a compost bin for horticulture; sheet welded into battery overspill boxes for lifeboats; clear bottles of 125, 250 and 500 ml capacity; curtain rails; and domestic window frames.

Until the 1950s, lead provided the best corrosion resistance in drainage pipes, with cast iron being the cheaper choice. The shortage and high price of lead then forced local authorities to seek alternative materials. Polyethylene

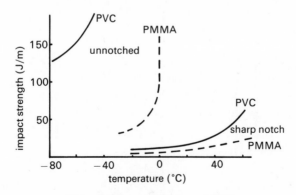

Figure 8.1 Impact behaviour of PVC

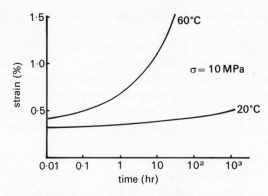

Figure 8.2 Comparison of creep of PVC at 20°C and 60°C

and UPVC were found to be not only suitable, but also offered several advantages: corrosion resistance, longitudinal flexibility, ease of installation, smooth internal bore, and economy. UPVC was preferred over polyethylene for drainage systems because of its rigidity and ease of jointing by solvent welding. BS 4514:1983 provides that pipes for domestic drainage have to withstand short exposures to water at 91°C and a rest period of 60 seconds, repeated for 2 500 cycles. In rainwater goods applications, UPVC is chosen because it combines rigidity, low cost, good UV stability, good chemical resistance, and ease of assembly and joining. The above-ground system is usually coloured grey, the below-ground system brown.

The below-ground system has to withstand the temperatures specified in BS 4514, but also the loads imposed by soil cover. The recommendations for such a pipe are a minimum wall thickness of 3.4 mm for a nominal inside diameter of 100 mm. BS 4660:1973 provides for all underground drainage systems to resist the crushing load of backfill. The grade of UPVC chosen for these applications must have a Vicat softening point not lower than 81°C, with 79°C at a 49 N load for fittings when tested to BS 2782: Part 1:1976 (1983), method 120B.

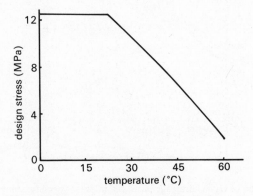

Figure 8.3 PVC: effect of temperature on design stress

Many companies have obtained Agrément Certificates for their own designs of underground drainage systems. UPVC has faced limited competition from other polymeric materials in these applications, but with the exception of drainage inspection chambers, UPVC has not been displaced. Drainage inspection chambers are complex units with several inlet and outlet ports which are injection-moulded in ABS or newer grades of HDPE, or fabricated from GRP. For pipes, UPVC has stood the test of time, having the best combination of properties – rigidity, chemical resistance, solvent welding and

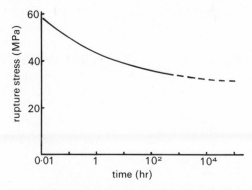

Figure 8.4 UPVC: creep rupture at 20°C

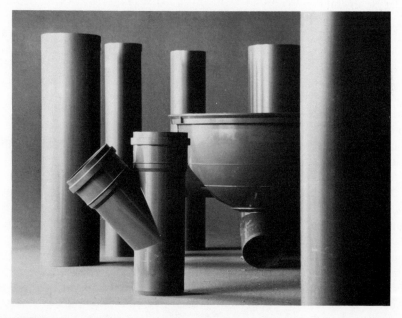

Figure 8.5 Selection of UPVC drainage pipes and fittings (Yorkshire Imperial Plastics) Photograph by courtesy of ICI Ltd

cost. The other cheap commodity plastics, polypropylene and polyethylene, generally are less rigid, difficult to join (although recent developments of on-site electrothermal welding, or friction-fit jointing have reduced this problem), and have poorer UV resistance, but have better chemical resistance.

BS 3505:1968 (1982) recommends a design stress for UPVC of 12.3 MPa at 23°C for cold-water service pipe (Figure 8.3). This standard requires the pipe to be in service for 50 years, corresponding to 30 MPa rupture stress (Figure 8.4). This gives a safety factor of about 2.1. However, as pipes in service will suffer from poor installation, external abrasive materials (e.g. broken bricks and stones), or simply poor manufacture, external and internal pressures will cause a pipe to fail at a stress much lower than 30 MPa. In critical applications it is preferable to design using fracture mechanics; indeed, a fracture toughness of UPVC is specified in BS 3505:1986. This results in a design stress of approximately 6 MPa. Figure 8.5 shows a selection of UPVC pipes and fittings.

8.2 Plasticized poly(vinyl chloride), (PPVC)

Plasticizers are high-boiling liquids which are solvents for PVC; increasing addition reduces the T_g value of PVC until the material is rubbery, usually at levels above 30 phr by weight. Most commonly used plasticizers are C_8 to C_{10} derivatives of phthalic acid, e.g. dioctyl phthalate (DOP) and didecyl phthalate (DDP): the latter is less water-extractable, hence finding use for baby-pants. Phosphate plasticizers (e.g. trixylyl phosphate) are more expensive, more toxic but effectively enhance fire resistance. Aliphatic acid esters with C_8 to C_{10} are normally used in combination with phthalate plasticizers.

Polymeric plasticizers, such as poly(propylene adipate) and poly(propylene sebacate), are non-volatile, non-migratory and resistant to hydrocarbon extraction. However, they are expensive and the compounds are difficult to manufacture. Extender plasticizers are not compatible with PVC but are miscible with plasticized PVC. A processor can use them to replace a part of the more expensive plasticizer content. Examples include chlorinated paraffin waxes, paraffin liquids or oil extracts.

Compounding of plasticizers, stabilizers and other additives into PVC is as critical as the subsequent shaping. Suspension-grade PVC is normally dry-blended using high shear mixers to produce powders for melt processing, e.g. calendering and extrusion. Emulsion-grade PVC compounded with levels of plasticizer above 20 phr tends to form free-flowing *pastes* (mixtures of plasticizer-swollen PVC particles suspended in the remaining liquid plasticizer – this will also be absorbed by the PVC, the rate depending on PVC resin, plasticizer, and temperature); these are also known as *plastisols*. A wide range of compounding equipment is used, from low-shear ribbon mixers to high-shear impeller mixers, preferably operating under vacuum to minimize air entrapment. Pastes lend themselves to spread, dip, screen and spray-coating, and the simpler shaping techniques such as pour casting and rotational moulding.

The controlled absorption of the plasticizer by the PVC particles is an essential part of the process, and this depends on particle size and texture. Paste viscosity depends on particle size and distribution. Paste storage life depends on plasticizer type (for instance, butyl benzyl phthalate, BBP, will produce rapid solvation and viscosity increase, compared with DOP), and the residual emulsifier left on the PVC particles: both these paste properties are dependent on its thermal history. Additives, such as inorganic fillers, will also have a marked effect on paste properties. Increasing the temperature of a paste increase plasticizer absorption and viscosity. Eventually all free plasticizer will be absorbed, with the swollen PVC particles in contact with each other and limited fusion occurring across particle boundaries. In this so-called *gelled* state, the plasticized PVC will be dry to the touch and have minimal strength. Further heating will increase the degree of *fusion* between the particles of PVC, until optimum mechanical properties result. The particle structure is very durable and indeed will be retained at process temperatures up to 170°C. Homogeneity or full fusion will occur above these temperatures as the PVC increasingly takes on the behaviour of a true thermoplastic melt, with relative movement within the molecular structure. Full fusion, indicated by maximum mechanical properties of a tough rubbery product, depends on heat input, previous thermal history, PVC type, plasticizer type and content, and other additives.

At lower additions of plasticizer (up to 10 phr), the expected softening behaviour is reversed and the compound increases in hardness and T_g. This *anti-plasticization* has been explained by increased crystallinity facilitated by plasticizer.

Other additives are used. Fillers serve both technical and economical purposes: china clay enhances electrical properties, but calcium carbonates and asbestos are also important. Pigments are used as coloured fillers: titanium dioxide acts as an intensive whitener and will reinforce the colouring effect of printing inks applied to the PVC substrate. Chemical blowing agents, such as azodicarbonamide (AC), break down to form gases during processing which expand PPVC into a cellular or foamed product. Cd stabilizers will also catalyze this breakdown, allowing the blowing action to be activated at lower temperatures. Because of the higher price of emulsion PVC, pastes are often made from resin blends including micro-suspension grades of PVC (up to 30%), known as extender resins; these also increase storage stability without sacrificing mechanical properties. It is important to note that increasing levels of phthalate plasticizers will reduce the fire resistance inherent in UPVC products.

Products made from PPVC based on suspension-grade polymer include calendered sheet; domestic electric cable; commercial damp courses; garden hoses made by the extrusion process; and automotive door seals and draught excluders made by the simultaneous co-extrusion of several grades of PPVC containing different levels of plasticizers. Figure 8.6 includes a few of the wide range of goods made from plasticized PVC.

Table 8.1 indicates the annual consumption of emulsion-grade PVC used in plastisol processes. The major use of emulsion-grade PVC has been in sheet

Figure 8.6 A selection of goods made from suspension, emulsion and copolymer grades of PVC. Permission of Wacker Chemicals (UK) Ltd

Table 8.1 Uses of emulsion-grade PVC in the UK, 1985. Permission of Wacker Chemicals (UK)

Applications	Approximate tonnage used
Sheet vinyl cushion flooring	28 000
Wall covering*	20 000
Metal cladding	10 000
Conveyor belting	7 000
Dip coating (e.g. gloves)	5 000
Car underseal	1 500
Total including miscellaneous applications:	80 000 to 85 000

*For 1986, it has been estimated that high profiled wall coverings based on foamed PPVC, probably used as much as 10 000 tonnes of PVC.

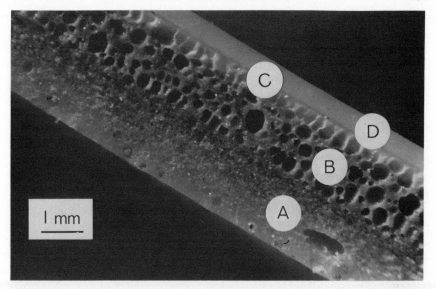

Figure 8.7 Micrograph showing the typical structure of a modern sheet vinyl flooring, made up of: (A) glass-fibre reinforced PPVC, (B) foamed PPVC, (C) decorative film based on printing inks, (D) clear PPVC wear layer. By permission of Armstrong World Industries Ltd

vinyl flooring, although structures have changed markedly over the last 20 years. Modern floor-covering is based on a composition of at least four laminates of diverse PPVC compositions, as illustrated in Figure 8.7.

Typical of many manufacturing methods, Armstrong's Rhinofloor™ sheet vinyl flooring is made up as follows. An inorganic particulate-filled paste is made up from phthalate and other plasticizers (approximately 60 phr) in a blend of PVC resins (e.g. $K = 70$ emulsion resin and up to 30 phr of extender suspension resin at $K = 55$ to 65). This is reversed-roll or knife-over-roll coated into an open glass-fibre (non-woven) scrim: the paste is *gelled* at 150°C to give limited mechanical strength. A foamable plastisol, containing a paste of AC blowing agent and titanium dioxide as a whitener, again based on a mixture of emulsion and suspension resins with fillers, is applied and gelled at 140°C without activating the blowing agent. The structure is rotogravure printed, applying up to five ink colours normally based on acrylates. Some of the ink will include an inhibitor to prevent the catalysis of the AC in the fusion process. The gelled PPVC should have enough strength to resist the pressure of printing. The structure goes back to the coating line for application of a clear wear layer. This must be transparent and be both abrasion- and stain-resistant; it is based on emulsion-grade PVC of relatively high K values, to give the required mechanical properties, in a plasticizer blend (e.g. DOP, BBP and secondary ester plasticizers to 60–70 phr total; the BBP is included to enhance stain resistance). Again the temperature of the oven is selected for gelation of the wear layer. Finally, either a solid or foamable back coat is applied to the back of the structure; the whole product is passed into an oven at 170–195°C, where fusion and foaming are completed. At the same time chemical embossing in-register is imparted by inhibiting the blowing action of the AC in selected parts of the flooring.

8.3 Vinyl chloride copolymers

Copolymers of vinyl chloride with vinyl acetate (VAC) prepared by granular polymerization are the most important. As crystallinity of the PVC is almost certainly eliminated by copolymerization, a typical copolymer with 13% VAC has a lower T_g and much improved flow properties, compared with the homopolymer.

Unplasticized copolymer is used for gramophone records; the chief demands for this product is quality of sound reproduction and reasonable cost. The four cheapest materials (polyethylene, polypropylene, polystyrene and PVC) could be considered, but polypropylene and polyethylene are both crystalline and exhibit larger shrinkage than amorphous materials. In addition, polyethylene tends to smooth out any sharp features (the record groove), and polypropylene, although it follows the mould surface, has high shrinkage; therefore neither is a suitable material for sound reproduction. Polystyrene is too brittle and has poor wear properties. Although the vinyl chloride homopolymer has some crystallinity, vinyl acetate in the copolymer inhibits this, giving an amorphous material which has good-quality reproduction, good wear properties, comparatively low cost and high durability. The copolymer is relatively cheaper to process than the homopolymer, having improved melt flow properties and a lower processing temperature.

The copolymer is blended with several additives to improve the performance of the record. For example, static charge can accumulate on the record surface, impairing the sound reproduction: antistatic agents such as quaternary ammonium salts have some limited value in solving this problem. Fillers cannot be tolerated in this application, although pigment is included, and stabilizers and lubricants are chosen to produce minimum surface noise. A typical formulation for records is:

	phr
Vinyl chloride/vinyl acetate copolymer (VC/VA, 87:13)	100
Dibasic lead stearate (lubricant/stabilizer)	0.75
Dibasic lead phthalate (stabilizer)	0.75
Lamp black (pigment)	2.0

Records are moulded using pre-heated extruded compound (dough) at 130–140°C; the cycle time varies from 20–30 seconds. It is possible to mould records in an injection/compression sequence on an injection machine. The machine operates in the conventional way, except that the mould is not completely closed (about 6 mm clearance), the melt is injected into the cavity space and then the mould is closed. However, the injection/compression method produces inferior records in respect of reproduction quality.

The main use of the vinyl chloride – vinyl acetate copolymer is in the production of flooring tiles. Typically, tiles contain 30–40 phr of plasticizer and 300–400 phr of filler (e.g. chalk, asbestos, pigments). The mix is dry-blended in an internal mixer and calendered.

8.4 Blends of poly(vinyl chloride)

PVC which has been processed competently, and where high stress concentrations are absent, has impact properties which are quite acceptable, as may be judged from the list of applications given above. However, there are designs for which increased toughness is demanded. PVC will form blends with a number of polymers, in both single and multiple combinations. The PVC is permanently plasticized by the other polymer(s) and the resulting material should ideally have a combination of the parent polymer properties, or better. Miscibility of any blend is not essential to produce property enhancement and is dependent on polymer types, their ratio and compounding method.

PVC with chlorinated polyethylene blends has improved impact performance, especially in the presence of notches and at lower temperatures than unmodified grades of PVC. They are used in continental Europe for external applications (e.g. rainwater goods); unmodified UPVC would be acceptable in the UK. Such blends are finding increasing use throughout Europe in extruded sections for window frames.

PVC blended with ABS or impact-modified PMMA has improved processability and good fire resistance, along with impact properties superior to those of the parent polymers. ABS/PVC is used as calendered sheet thermoformed into car crash-pad skins (subsequently back-filled with semi-rigid PU foam), also in moulded power tool handles, sanitaryware and electrical housings. Impact PMMA/PVC is used for seat backs for public transport and industrial panelling.

PPVC is blended with high-molecular-weight elastomeric materials, such as nitrile rubber (NBR), EVAC and polyurethane (PU), to improve the shock resistance of flexible compounds. PPVC/NBR blends have good oil resistance and are used to make industrial boot-soles and in some electrical cable applications. PVC/EVAC and PVC/PU are used in shoe-sole formulations.

8.5 Vinylidene chloride polymers and copolymers

A considerable number of other chlorinated polymers find commercial use as a means to impart fire retardancy into a product. Poly(vinylidene chloride) (PVDC) and its copolymers are described here, particularly because of their gas and vapour barrier properties.

Vinylidene chloride is produced by the dehydrochlorination of trichloroethane at 400°C. The monomer is difficult to handle because of its flammability, toxicity and its readiness to autopolymerize. The monomer is normally suspension polymerized using peroxide catalysts at 0–20°C: emulsion and bulk methods are also used. The polymer has a regular molecular structure, giving rise to high crystallinity, high specific gravity (1.875) and low permeability to gases and vapours (up to a thousand times better than LDPE). The high chlorine content in the polymer produces excellent fire resistance but also leads to rapid degradation at normal process temperatures.

Copolymerization by suspension or emulsion methods, with vinyl chloride or acrylonitrile for example, reduces the regularity in the chain, increasing flexibility and permitting processing below the copolymer's degradation temperature. Copolymers are readily crystallizable, and care is needed in the setting of process temperatures for extruded (e.g. pipe) and moulded goods.

The polymer and its copolymers are mainly exploited for their barrier properties, both as film and coating materials. The major use for homopolymer and vinyl chloride copolymer is a surface coating for PETP bottles, to enhance their oxygen barrier properties, when the bottle is to be filled with a carbonated alcoholic beverage: alcohol may be oxidized to acetaldehyde, which taints the flavour of the drink. The outer surface of the bottle is dip-coated in polymer as a latex, and dried at about 60°C (i.e. below the T_g of PETP), to form a bonded, coherent film on its surface. The biaxially oriented film has excellent mechanical and optical properties. The copolymer is also made into fibres which have excellent mechanical properties, durability and chemical resistance (they are used in deck-chair fabrics for example).

The acrylonitrile copolymers find limited use in the production of cast film for coated cellophane, polyethylene and paper.

9 Speciality thermoplastics

A number of thermoplastics produced in small quantities worldwide have already been described within the appropriate families of materials in earlier chapters, for example polyamides and polyimides. The present chapter accommodates the more important members of the thermoplastic groups not included in other sections. These include fluoroplastics, polysulphones and sulphide polymers.

9.1 Fluoroplastics

Poly(tetrafluoroethylene) (PTFE) is the oldest member of this family, having been available since the early 1950s, yet worldwide production is only a few thousand tonnes per year. Even so, PTFE is the most commonly used fluoropolymer, widely used in specialist applications by engineers. This is primarily because of its unusual properties, notably high melting temperature, extreme chemical inertness and an extremely low coefficient of friction for either PTFE–metal or PTFE–PTFE contact. On the debit side, PTFE has an inherently high price and is very difficult to process, which is principally associated with the extremely high viscosity of the molten polymer. Many attempts have been made to improve the processing behaviour, mostly by copolymerization, without sacrificing too many of its desirable characteristics. A limiting compromise is to substitute other fluorine-containing homopolymers such as poly(vinylidene fluoride), into part of the PTFE molecular structure; these do not present processing problems, but unfortunately they do not have the very desirable properties of PTFE.

Poly(vinyl fluoride) is included, although it shows greater similarities to poly(vinyl chloride) than to the other fluoropolymers. It also should be mentioned that a wide variety of fluoropolymer elastomers is used under extreme conditions of temperature and chemical environment.

Fluorine-free polymers find particular use in high-temperature and corrosion-resistance environments, and some consideration will be given to them in this chapter.

9.1.1 Poly(tetrafluoroethylene)

There is a large number of companies manufacturing PTFE: this is surprising when the difficulties and hazards associated with the industrial preparation of the monomer and its polymerization are taken into account. Suspension and emulsion polymerization are employed to give materials suited to particular processing methods. Further, to obtain optimum properties in the final product, work-up operations for the polymer must be carried out under rigorously dust-free conditions.

Experimental and theoretical evidence indicates that suspension poly-merized (granular) PTFE is a linear and unbranched polymer, with a molecular weight exceeding half a million. PTFE has the comparatively high specific gravity of 2.28, and the virgin polymer has very high crystallinity, 90–95%, and a melting point of 340°C. However, after melt processing the crystallinity is not fully regained, dropping to between 50–70% with a melting temperature of 328°C. A reversible volume increase of about 30% takes place between 15°C and the melting point. Above the crystalline melting point, the 'melt' has a very high viscosity, typically a million times greater than for most thermoplastics: this precludes normal thermoplastic processing methods.

Lower crystallinity, and consequently density, are associated with higher-molecular-weight polymer. This is a useful relationship, since the normal solution methods of molecular weight determination are precluded by the insolubility of PTFE in all solvents except at temperatures above 300°C, which property is exploited in many applications. PTFE is chemically inert, attack being limited to materials such as molten sodium. It may be dissolved by highly fluorinated liquids near its melting point. It has excellent weathering resistance and is stable in air to at least 250°C for continuous service.

Very low coefficients of friction are obtained for PTFE–PTFE or PTFE with most metal systems: as a corollary, nothing will stick to PFTE. The low critical surface tension of PTFE ($18 \, mNm^{-2}$) prevents wetting and subsequent adhesion by adhesives or surface coatings, unless the PTFE is pretreated (e.g. by plasma or sodium xanthate etching). Dielectric losses are very small, particularly for unsintered material, and volume resistivity is very high. Mechanically PTFE is tough and moderately flexible with a high elongation to break. Toughness is retained at very high and very low temperatures: PFTE is one of the few materials which does not embrittle at liquid-helium temperatures. On the other hand, PTFE creeps under moderate stress, a design consideration and possible limitation on the engineering use of the material. Recovery from deformation is poorer than for other thermoplastics. Many applications of PTFE involve compression (see Figure 9.1), and therefore compression creep is important: representative creep data are given in Figure 9.2.

Although PTFE is a linear-chain polymer and a thermoplastic, its extremely high melt viscosity does not permit the usual thermoplastic processing technology. Instead the methods of powder metallurgy are adopted: pressure preforms are sintered at temperatures above the crystalline melting point, typically 380°C. Complicated shapes are obtained by machining, making products very expensive. Powder processing is a compromise between powder

Figure 9.1 PTFE expansion bellows: weight 7 kg. Made by Henry Crossley (Packings) Ltd., Bolton, from Fluon™. Photograph by courtesy of ICI Ltd

flow and good strength; particle shape, size, size distribution, external topography and internal structure are all important. Some highly toxic cyclic fluoro-compounds are liberated by heating PTFE to sintering temperatures, so good ventilation is essential. Extrusion, particularly of thin sections, is accomplished by lubricating dispersion PTFE powder with petroleum ether, and subsequently drying and sintering the extrudate (tube for example). Unsintered tape is obtained by calendering.

Applications cover nearly all property regions. PTFE is exploited for non-stick coatings in a variety of applications, including cookware, bread processing, toffee-making and coating on skis. Bearings, particularly composite systems, frequently involve PTFE. An exacting application is in suspension bushes used in some rally cars, which have been spectacularly successful. Chemical inertness in the presence of biological fluids is one basis for the many medical applications of PTFE: a typical use is in the catheters used in dialysis equipment employed in the treatment of renal failure. PTFE is widely used for critical electrical insulation, but polyimides and polyketones

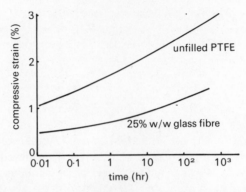

Figure 9.2 Compressive creep curves for PTFE: 20°C, 6.9 MPa

are now preferred for wire coating applications. The unsintered, extended material is used by plumbers, as tape for sealing threads. Valves, gaskets, expansion bellows are typical uses for PTFE. An example is an expansion bellows (Figure 9.1), in a glass-reinforced polyester pipeline carrying wet chlorine gas. In a large chlorine-producing plant, chlorine is piped to the cleaning and drying units together with brine mist at low pressures (14 kPa or 2 psi) and at a temperature of 70–80°C. Transporting this corrosive mixture presents several problems, the greatest of which is to provide corrosion-resistant pipework of long life. Furthermore, the pipework should allow for expansion and contraction with changes in temperatures. Polyester resin was chosen for the pipework but the existing rubber-based expansion bellows had a short service life, only 12–18 months. PTFE was recommended for this application and ensures that the bellows have a virtually unlimited life. As its reliability means no lost production, the bellows are much more cost-effective than rubber, although much more expensive initially.

Many of the properties of PTFE depend on the characteristics of the particles (which can be affected by polymerization technique) and on the processing conditions, including moulding pressure, sintering temperature and time, and heating and cooling rates. Crystallinity and porosity are affected, which in turn influence properties such as strength, modulus, hardness and permeability. Fillers are used in PTFE to give particular technical effects rather than, as might be expected, to reduce the cost. Creep behaviour is improved by inclusion of short glass fibres (Figure 9.2); the excellent frictional characteristics are further enhanced by incorporating graphite or molybdenum disulphide, or both; metal powders increase thermal conductivity thus reducing overheating and consequent failure in bearings.

9.1.2 *Poly(vinylidene fluoride)*

Poly(vinylidene fluoride) (PVDF) is a fluoropolymer which shares some properties, including high cost, with PTFE. It contains the repeat unit

—CH$_2$—CF$_2$—, explaining its chemical inertness, but it will degrade at high temperatures (380–400°C), liberating HF. It has a crystalline melting point of 170–171°C. The various grades of PVDF homopolymer are available for most thermoplastics processing methods without significant degradation at 220–240°C. PVDF is incompatible with other polymers (except acrylics) and the melt may be degraded by impurities. Therefore process equipment must be scrupulously clean and care must be exercised in the selection of pigments, although PVDF may be recycled at normal melt temperatures.

The properties of PVDF are influenced by its molecular weight, which also strongly affects processing and the crystallinity level achieved. Generally, the properties are not strongly dependent on operating temperature in the range 60–150°C; this leads to applications in chemical plant, transport and aerospace. In such applications as pipes, pumps and fittings, poly(vinylidene fluoride) is resistant to halogens, oxidants, strong bases and acids (except at high temperatures); it is less resistant to polar organic solvents. One of its main application areas is in chemical process plant, PVDF having good strength and impact resistance and outstanding resistance to abrasion/erosion. This exceptional property has led to the satisfactory use of PDVF as a coating on propellers for high-performance water craft. Representative creep data are given in Figure 9.3.

The resistance of PVDF to γ-radiation is exceptional; the material is little affected even by a dose of 300 Mrad from a Co^{60} source. In the reprocessing of nuclear waste, reliability of the equipment is of the utmost importance. Reliability has to be maintained during the handling of corrosive liquids such as nitric and hydrofluoric acids at 105–110°C, while exposure to high-energy radiations is also a major consideration in this application. On all these counts, PVDF shows satisfactory behaviour, making it a recommended material for pipework in this exacting application.

Its resistance to ageing and weathering is excellent, and so it finds ready use in polluted industrial locations: eight years' exposure in such an environment

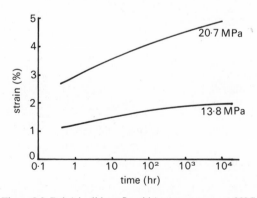

Figure 9.3 Poly(vinylidene fluoride): creep curves at 20°C

is one example resulted in only marginal change in mechanical properties. PVDF is also self-extinguishing.

Filled compositions are available, but it must be remembered that fillers can reduce the stability of the polymer or introduce instability in their own right, glass beads reduce stability in the presence of hydrogen fluoride, for instance. Filled compounds show markedly lower shrinkage and higher rigidity; toughness is generally decreased.

Copolymers of vinylidene fluoride and tetrafluoroethylene are available, characterized by lower crystallinity and melting point, with greater flexibility, extensibility and toughness. Otherwise they are generally similar to the homopolymers.

9.1.3 Other fluoropolymers

Polyvinyl fluoride (PVF) is generally similar to PVC, but is transparent to UV radiation and therefore has excellent weathering resistance. It is used in laminates because of this property. Its brittle temperature is very low, $-165°C$.

Polychlorotrifluoroethylene, (PCTFE) is less crystalline and less inert than PTFE: it can be shaped by normal thermoplastic melt processing methods. Because of its relative transparency, it finds limited use as observation windows (in sheet form), or as specialist tube material.

Other fluoropolymers are produced in small quantities and for specific applications, including copolymers of ethylene with chlorotrifluoroethylene and TFE.

9.2 Polysulphones

These plastics are characterized by having the sulphone group $—SO_2—$ as a constituent in a polymer structure consisting largely of aromatic groups, generally linked in the *para*-position (*p*-phenylene). Thermal stability in air, high softening temperature and good mechanical properties are attributes leading to an increasing interest in a variety of applications, in spite of their high cost. Polysulphones can be readily processed by conventional thermoplastic techniques although their softening temperatures exceed 200°C.

Polysulphones may be formally expected to have regular and linear structures, but in fact most types contain a 'kink' or 'dog-leg' which inhibits crystallization. This similarity exists in the polycarbonates, and it is presumed that the reluctance to crystallize is related to similar structural features.

(i) Polycarbonate

(ii) polysulphone, e.g. Udel™ Polysulfone (Union Carbide), softening point 190°C

(iii) Polyethersulphone (softening point 220°C)

Therefore, most commercial polysulphones are available as amorphous materials, which consequently are transparent after melt processing. There is an exception in the family, which does not possess the 'kinked' feature (iv); this is a crystalline polymer having a melting point exceeding 500°C, at which temperature it will be also starting to degrade:

(iv)

Crystallinity is not a characteristic of the commercially significant polysulphones, which consist of the homopolymers (ii) and (iii), together with a range of copolymers based on (iii) and (iv). The inclusion of increasing amounts of the latter increases the softening point of the polyethersulphone.

Exemplifying the design properties of these materials by reference to data for polyethersulphone, creep data for unmodified and glass fibre-reinforced grades are given in Figures 9.4 and 9.5 for 20°C and 150°C, and for strength as a function of temperature in Figure 9.6. Impact data indicate that the material

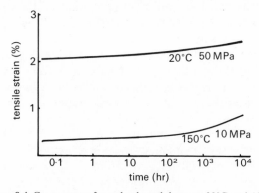

Figure 9.4 Creep curves for polyethersulphone at 20°C and 150°C

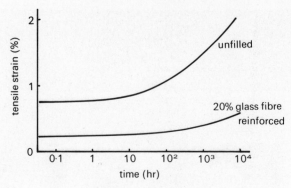

Figure 9.5 Tensile creep of polyethersulphone at 150°C, 20 MPa: effect of glass fibre reinforcement

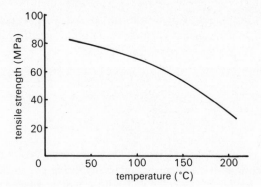

Figure 9.6 Tensile strength of polyethersulphone

is tough at 20°C in the absence of stress concentrations; with an Izod notch, the impact strength is 48 Jm^{-1}. This plastic possesses outstanding loadbearing properties at normal and elevated temperatures.

Polyethersulphones can be used for tens of thousands of hours at 200°C without significant loss in strength, a property which is the basis for a number of applications, including internally illuminated push-button switches and indicator lights for cookers. This excellent stability at high temperatures is not matched by ultraviolet resistance in weathering, which is poor. However, resistance to ionizing radiations, (β-, γ- and X-ray) is good in the temperature range 20–200°C. Polyethersulphone is a highly polar aromatic polymer, thus, as expected, it is resistant to aqueous acids, bases and most salts and to hydrocarbons, including petrol, oils and greases, even at high temperatures, where glass-filled grades are recommended. A typical application is in the inlet manifold between carburettor and cylinder in a petrol-engined power chain-saw. Polar solvents, particularly of an aromatic type, are liable to dissolve polyethersulphone partially or completely. In the presence of residual or externally applied stresses, polar solvents will also give rise to crazing.

Figure 9.7 Polyethersulphone: element housing that locates the heater element of Anda high speed hand drier, moulded by Wells and Hingley Ltd., Catford. Photograph by courtesy of ICI Ltd

Figure 9.8 Polyethersulphone integrated circuit socket developed and moulded by Jermyn Industries Ltd., Sevenoaks

Polyethersulphone is used in the heater element holder of a high-speed hand drier (Figure 9.7). This unit has to endure a working temperature of 100°C, with peaking temperature under fault conditions of 200°C, or even higher. Polyethersulphone is eminently suitable for these temperature conditions and, since it is unlikely to encounter solvents in this application, it would be expected to have a very long service life. The part can be manufactured by injection moulding on conventional equipment and has in fact, proved successful in service. The non-flammability of the polymer is relevant in this application, a characteristic leading to its use in aerospace applications where cost considerations are minor compared to safety factors. A further attribute of polyethersulphone is its ability to be shaped to fine tolerances, which is also dependent on its inherent dimensional stability. An application illustrative of this property is as a base fitting for integrated circuits (see Figure 9.8).

9.3 Poly(phenylene sulphide), (PPS)

Poly(phenylene sulphide) has a symmetrical repeat unit consisting alternately of sulphide links and *para*-phenylene groups, so it is not surprising that this is a crystallizable plastic.

$$\sim\!\!\!\langle\!\!\bigcirc\!\!\rangle\!\!-\mathrm{S}\!\sim$$

The melting point, 277°C, is not particularly high, considering the predominance of the aromatic groups in the structure (compared with polyamide 6.6 with $T_m = 265°C$). The T_g of the polymer (88°C) is marginally higher than that of poly(ethylene terephthalate), 70°C, suggesting that some problems might arise in processing if rapid chilling is imposed. Quenching does indeed affect the crystallinity significantly, as shown in Figure 9.9, which depicts the flexural modulus of 'annealed' and 'as-moulded' specimens of

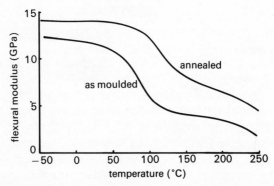

Figure 9.9 Flexural modulus vs. temperature for 'as moulded' and 'annealed', 40% glass fibre-filled PPS

poly(phenylene sulphide) containing 40% glass fibre. Unfilled PPS is not important commercially, the significant materials being those containing glass fibre or combinations of glass fibre and mineral fillers; the latter compounds require drying prior to processing because of the poor hydrolytic stability at process temperatures.

Processing is usually by injection moulding for which the melt temperature is 315–360°C. Mould temperatures of 50–160°C have been used, although temperatures above 130°C are preferred, giving increased crystallinity, better dimensional stability at elevated temperatures, and improved gloss. High filler content gives rise to low mould shrinkage (0.2–0.6%) for a crystallizable polymer.

PPS has been called 'the poor man's polysulphone' because of its relative cheapness; this refers to some similarity in properties between the polymers, particularly temperature resistance. There is a significant change in the mechanical properties of PPS, as shown in Figure 9.9, associated with a glass transition temperature below 100°C, a phenomenon which does not occur with polysulphones. The resistance of PPS to high-energy ionizing radiations is excellent, probably surpassing that of polysulphones. However, its poor resistance to UV radiation encountered in weathering is similar to that of polysulphones: fortunately this behaviour can be markedly improved by the addition of stabilizers. On the other hand, the chemical resistance of PPS is inferior to that of the polysulphones, particularly to oxidizing solutions, chlorine and chlorinated hydrocarbons and amines, although PPS is a crystalline plastic. It may be inferred that the sulphide linkage, ($-S-$) is not as chemically inert as is the sulphone group ($-SO_2-$).

An outstanding characteristic of PPS is its resistance to fire. It is self-extinguishing (having a critical oxygen index of 50), although if a flame is present PPS will burn, charring without dripping or fire-spread. At the same time, some smoke may be formed, but the concentration of potentially toxic gases (e.g. carbon monoxide and mercaptans) is low.

Typifying the use of PPS is the handle of a carpet-bonding iron (see Figure 9.10a). A similarity between cross-linked phenolics and poly(phenylene sulphide) is that both can withstand operating temperatures up to 250°C: Table 9.1 reproduces some of their properties.

These two materials differ considerably in mechanical properties, however: the cheaper phenolics have much lower tensile strength, along with low tensile elongation (0.5%). Both have good electrical properties and will withstand direct contact with metal at 250°C. The main reason behind the choice of specific materials for these applications is the design of each part. The design with the open end (see Figure 9.10a) will have to withstand large flexural stresses at its corners. Also, if any impact loads on the cantilever end will induce large stresses in the component. Poly(phenylene sulphide) is therefore suited for this application because of its excellent mechanical properties and its suitability for operating at high temperatures (up to 250°C): by contrast,

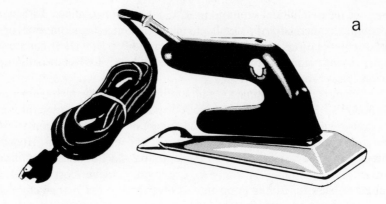

a

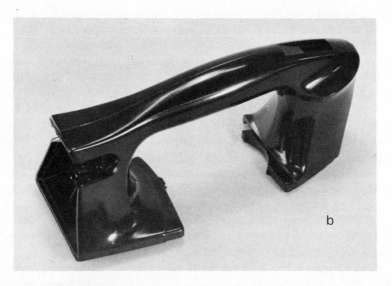

b

Figure 9.10 Designs of handles for irons: (*a*) poly(phenylene sulphide) (Roberts Consolidated Industries, California) (*b*) phenol-formaldehyde (Healey Mouldings, Oldbury)

Table 9.1: Properties of poly(phenylene sulphide) and cross-linked phenolics

	PPS	Phenolics
Cost	High	Low
Tensile strength	140 MPa at 100°C	50 MPa at 20°C
	20 MPa at 240°C	20 MPa at 250°C
Flexural modulus	12 GPa at 100°C	21 GPa at 20°C
	3 GPa at 250°C	
Impact strength	80 J,m^{-1} at 210°C	20 J,m^{-1} at 20°C
Deflection temperature	260°C	182°C

phenol-formaldehyde lacks the strength. Another feature in favour of the PPS is its high impact strength, which could be significant for equipment designed for industrial use. The enclosed handle (see Figure 9.10 b), has smaller bending moments at its corners when in use and so the stresses will be less. Furthermore the 'integrity' of the design means that it is inherently much stronger. Therefore both materials could be used in this application; however, the much cheaper phenolic is chosen since it will operate adequately under the applied stresses. The designer's task is to produce an attractive and useful product at an economical price. In designing the irons, one designer has tried to change the concept of the iron by adopting a cantilever shape, which should be technically more satisfactory in use. By doing so, employing a higher-cost material has resulted in a more expensive product. Nevertheless, the manufacturer has decided that a more attractive product with a price disadvantage would be more marketable than an iron with a more traditional design.

10   Cross-linked plastics

10.1 Introduction

A number of useful cross-linkable or *thermosetting* plastics are available. When cross-linked, these materials tend to be stiff, but with poor impact strength and low elongation. Normally reinforcing fillers have to be incorporated into the polymers to produce strong, useful materials, known as *composites*. Selected fillers can be also used to enhance electrical properties, such as volume resistivity, or to reduce the cost of the compounded polymers. Cross-linked polymers used in large volumes include:

(i) *Phenol-formaldehydes* (PF), including the *novolak* moulding powders and *resol* liquid impregnation resins
(ii) *Amino-formaldehydes*, including urea-formaldehydes (UF) and melamine formaldehydes (MF), (also known as *amino-plasts*)
(iii) *Unsaturated polyesters*, glass-fibre impregnation resins, dough-moulding and sheet-moulding compounds (DMC and SMC respectively), and moulding powders
(iv) *Epoxide resins*, normally available as liquid impregnation resins.

Other thermosetting materials are used in relatively small quantities in specialist applications, including alkyd and silicon resins; polyurethane resins are discussed in the next chapter.

Thermosets differ fundamentally from the thermoplastics in many respects. To form a cross-linked polymer, the reactants used will have functionalities greater than two. It is important to note that the shaping stage is associated with chemical change in the system. Since chemical reaction is inevitable in processing, the opportunity can be taken to reduce many of the difficulties found with thermoplastics shaping. In particular, the very high melt viscosity inherent with thermoplastics is avoided by working with comparatively low molecular weights and reactive *prepolymers* (or *oligomers*). Prepolymers exhibit certain thermoplastic characteristics, but will melt at relatively low temperatures to give low-viscosity liquids—some are mobile liquids at room temperature, or are made so by the use of solvents, such as unsaturated polyesters dissolved in styrene. With a few exceptions, very high-pressure moulding equipment is not required, and shaping of thermosetting materials can be accomplished on relatively low-cost equipment. Accurate proportioning and mixing equipment is needed to handle dual-component systems.

Fillers may have a marked effect on the flow properties of these materials.

As processing is associated with chemical changes in the polymer, there are several interactions of the prepolymer with the shaping operation. Prepolymers can either react with themselves, or by addition of a reactive cross-linking or *curing* agent. The reaction proceeds initially by chain extension with an increase in molecular weight, and then cross-linking to form a three-dimensional network system, which will eventually, in theory, have an infinite molecular weight. Both addition of catalysts and increasing the temperature of the reactants will increase the reaction rate. So, in any process, heating to effect melting and flow must not bring on premature cross-linking before shaping is complete. As the chemical changes are associated with increased molecular order, appreciable shrinkage frequently occurs on curing. In practice, this can be reduced by the incorporation of fillers, or of shrink or low-profile additives: these are discussed in detail in section 10.5.

A difference between the shaping of thermoplastics and that of thermosetting plastics is that the time-dominant cooling stage necessary to stabilize the shape of a thermoplastic component is replaced by the time-dominating cross-linking reaction taking place in a hot mould or press. Furthermore, while it is good practice in thermoplastics injection moulding to maintain approximately constant cross-sectional areas, and preferably constant thickness in the flow path, such restrictions are not so important for thermoset moulding.

There are two more advantages of thermosetting plastics. First, the absence of post-moulding shrinkage ensures that very close tolerances can be held, leading to acceptance for precision mouldings. Second, the low viscosity of the prepolymer at the shaping stage means that gates in injection moulds can be positioned as convenient, without so much consideration of features such as weld lines (although highly-filled compounds can cause problems).

10.2 Phenol-formaldehyde plastics

Phenol-formaldehyde (PF) resins were the first man-made plastics to be used, being first exploited commercially in 1910. Early applications were in the electrical industry where much of the development and manufacture was accomplished.

Phenol is a petrochemical derived from benzene and propylene by the cumene route, and formaldehyde is obtained by the oxidation of methanol which in turn is derived from synthesis gas. Until the 1950s, phenol, as extracted from coal tar by distillation, was contaminated with cresols and xylenols.

The PF prepolymers may be of two types, the *novolaks* and *resols*. The novolaks are prepared from excess phenol and formaldehyde (in the molar ratio 5:4) under acidic conditions. The monomers react slowly to form *o*- and *p*-methylol phenols (Reaction 1), which then condense rapidly with further phenol to give dihydroxyl diphenyl methane types (Reaction 2):

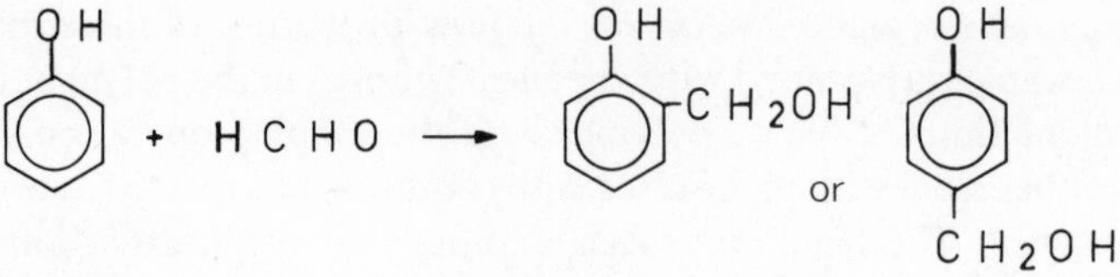

Reaction 1

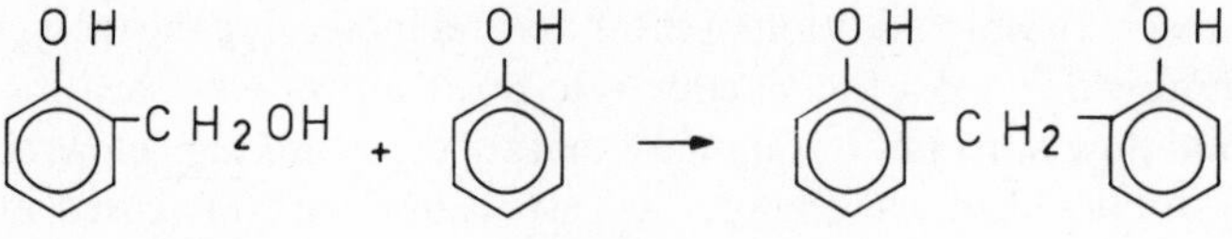

Reaction 2

Of the three isomers, 2.4′ and 4.4′ predominate. These then react to form methylol derivatives which in turn react with more phenol to give polynuclear phenols. Stoichiometry demands that the reaction stops when there are approximately five benzene groups per molecule (Reaction 3):

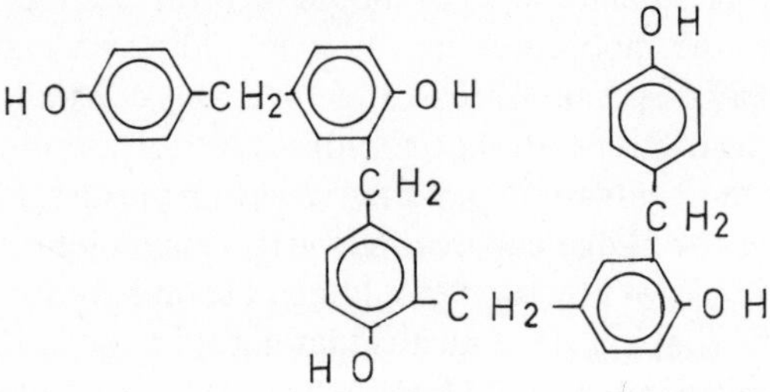

Reaction 3

The resulting novolak resins do not contain reactive methylol groups and so will not self-cross-link on heating. Instead, novolaks are compounded with reactants capable of forming methylene bridges at elevated cure temperatures and usually under alkaline conditions, e.g. hexamethylene tetramine ('hexa'), or paraform. A novolak resin is predominantly tri-functional, being open to cross-link formation at the o- and p-positions.

A resol is formed by reacting a phenol with excess of formaldehyde (phenol to formaldehyde ratio up to 1:2.5), normally under basic conditions. Phenol alcohols are formed rapidly, but their subsequent condensation is slow: thus there is a tendency for a mixture of phenol alcohols and methylols to be formed (Reaction 4). Heating of the resins will cause cross-linking via methylol groups or more complex routes. Novolaks and resols are soluble and fusible low-molecular-weight products, and cross-linking can be effected at temperatures below 160°C, occurring via the $-CH_2-$ and $-CH_2-O-CH_2-$ bridges. Above 160°C, quinone methides are formed (Reaction 5): these are conjugated double bond systems and account for the dark colour of PF plastics.

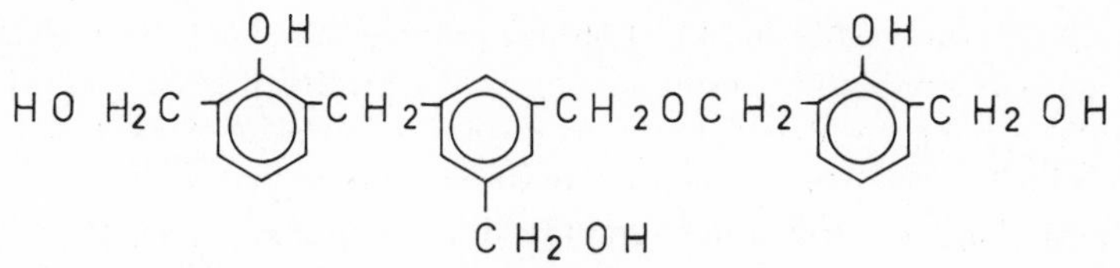

$$HO\ H_2C \overset{OH}{\underset{CH_2OH}{\bigodot}} CH_2 \bigodot CH_2OCH_2 \overset{OH}{\bigodot} CH_2\ OH$$

Reaction 4

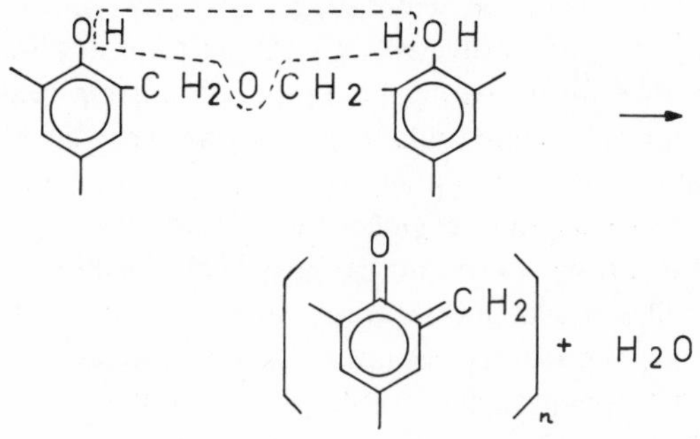

$$\overset{OH}{\bigodot} CH_2OCH_2 \overset{H\ OH}{\bigodot} \longrightarrow$$

$$\left[\overset{O}{\bigodot} =CH_2 \right]_n + H_2O$$

Reaction 5

Condensation products will be evolved during the cross-linking reactions and it is still a matter of debate what happens to them. It is assumed that, under the moulding conditions employed, they will form solid solutions in the cross-linked resin.

10.2.1 Novolak PF resins

Moulding powders are usually based on novolaks and 'hexa'. It is necessary to incorporate fillers to lower the resin content and reduce the shrinkage. Wood-flour is commonly used to improve impact strength and as a cheapener. For better impact strength, fibrous fillers (cotton flock, chopped fabric) are used; mica and asbestos are employed to give the best electrical properties. Improved strength is derived from covalent bonding at the interface between the PF resin and the cellulosic fillers. Curing is carried out at high temperatures for short cycle times, although dark products result.

Bulk applications of PF resins utilize novolaks in formulations designed for rapid cure. Traditionally, components were manufactured by compression or transfer moulding, the latter being preferred for products with moulded-in inserts. Injection moulding is gaining in use, for example for electrical components, electric iron handles (see Chapter 9), or saucepan handles (see below).

Since the final polymer is cross-linked and highly interlocked, phenolic mouldings are hard, insoluble, heat-resistant materials. Chemical resistance depends on filler and resin; PFs are attacked by caustic soda solution, but plastics based on cresols and xylenols are more stable. They are resistant to many acids except formic and oxidizing acids. Electrical properties are not outstanding, but are adequate for many insulation purposes. Compared to the amino-plastics, PFs have poor tracking resistance.

PF mouldings are still widely employed, for example for knobs, handles, instrument cases, bottle caps and closures. Although still used in some electrical equipment, particularly where heat- and moisture-resistance are important, PF has been largely superseded by urea- or melamine-formaldehyde, unsaturated polyester or epoxide resin in plugs and switches. PF is used in several automotive applications. It is still dominant in distributor heads and ventilation systems where good long-term heat resistance in a hostile environment is required: there is some competition from some of the higher-temperature-resistant engineering thermoplastics. Asbestos-filled grades are used in brake-shoes and exhaust heat shielding.

Saucepan handles are common components found in everyday use, for which plastics are eminently suitable in respect of their low thermal conductivity. Plastics can be easily shaped to complex sections which allow safer handling by providing a surer grip. Mechanical property requirements are very demanding because of the danger of failure in service, which could subject the user to scalding or worse. For example, a danger with a gas hob is that a flame can extend up the outside of the pan, subjecting the handle to softening heat, the possibility of ignition, and cracking with loss of mechanical strength. PF moulding materials have high rigidity and mechanical stability at high temperature coupled with flame resistance and low cost; in addition, the polymer can be cost-effectively injection moulded in large numbers (Figure 10.1).

Two examples of applications involving mechanical and electrical properties are the brush box moulding and a component of a multi-function electrical switch. The brush box moulding is a component of an electrical starter motor manufactured by Lucas Electrical Ltd. (Figure 10.2). The main purpose of the moulding is to locate and position the carbon bushes which make contact with the slip ring on the motor armature shaft. In order for the motor to operate efficiently, exacting tolerances are required in the position and the shape of the brush locations. Further, the product has to withstand localized hot spots in the areas around the brushes and must have good electrical properties.

The material chosen for this application was a short-fibre asbestos-filled phenolic because of its good electrical properties, resistance to localized high temperatures and close post-moulding dimensions which are achieved with very little moulding distortion. Originally the components were made using compression moulding in a four-cavity mould. The overall cycle time was 165 seconds, of which 120 seconds was the cure time. However, difficulties were experienced because the tool design was complex, and flow problems occurred using pre-heated pellets. If the pre-heated pellets were too hot when placed

Figure 10.1 Sculptured saucepan handles injection-moulded by Healey Mouldings Ltd, Oldbury, in phenol-formaldehyde compound

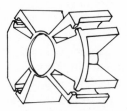

Figure 10.2 Brush box injection moulded by Lucas Electrical Ltd, from asbestos-filled phenol-formaldehyde

into the mould cavity, the PF would cure prematurely, producing discontinuity in the moulding; if too cold, the pellets would exert excessive force on the thinner sections of the mould, causing distortion.

As a result of the necessity to maintain and repair the compression moulds, injection moulding was considered as an alternative. The initial mould was

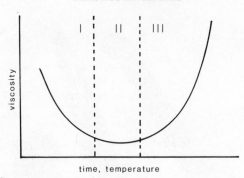

Figure 10.3 Flow-cure relationship for a typical phenol-formaldehyde moulding compound
      I Plasticization period
     II Injection stage
    III Curing process

designed using a three-plate concept, with a central feed supplying two cavities. Special attention was paid to the design of the runners, to ensure that the sprue and runners were separated from the moulding when the mould was opened. Injection moulding of thermosets requires accurate control of the moulding parameters, temperature, pressure, time and screw rotation speed, because of the flow behaviour of these materials. It is therefore necessary to understand the interrelationship between changes in viscosity with time and temperature (Figure 10.3). When the material is injected into the cavity it is essential that its viscosity is at a minimum to permit easy flow. Adjustments to the injection speed and barrel temperatures can be made until optimum mouldings are obtained, then the cure time can be reduced; the cure time will be the minimum time in which mouldings can be produced, without blistering or distortion.

During proving trials using injection moulding equipment with a suitable PF moulding powder, the minimum cure time was established at 45 seconds with an overall cycle time of 58 seconds. As a general rule, injection moulding of thermosets is much quicker than compression moulding because the material is much further advanced into its cure when the cavity is filled. In this example, a single injection moulding machine using two cavities would produce 42% more throughput than single compression moulding press with four cavities. However, this improvement has to be offset against the increase in waste material produced as sprue and runners; the overall shot weight was 43.5 g against the original 30.5 g by compression moulding. To reduce this wastage, warm runners were considered, but further difficulties were met: the mould had to be kept at the cure temperature (160–180°C) and the runners at the plasticizing temperature (60–80°C). Initially the runners restricted the flow, but satisfactory mouldings were produced by increasing their cross-sectional area. Ideally, the shot weight should equal the trimmed moulding weight, which is difficult to achieve in practice. However, after proving trials,

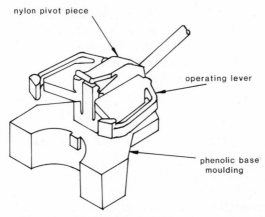

nylon pivot piece

operating lever

phenolic base moulding

Figure 10.4 Phenolic component in a multi-function switch for motor vehicles manufactured by Lucas Electrical Ltd

a stalk 20 mm long remained connected to the moulding, with the final and acceptable shot weight of 34.5 g.

The base moulding shown in Figure 10.4 is a component of a column-mounted multi-function switch used in motor vehicles to operate the direction indicators and horns, and to select main or dip beam. The base is required to bolt to the steering column, to locate the rotor and act as a bearing for the rotation of the rotor, while holding the contacts for the switching of the direction indicators.

The first prototypes were made with both base and rotor made from polyacetal. The unlubricated acetal mating surfaces of the switch 'squeaked' in operation, giving rise to an undesirable and annoying side effect. Lubrication was not acceptable because of the close proximity to the electrical contacts. The alternatives were either to place another material between the rubbing surfaces, or to change the polymer. A polyamide washer was tried and, although successful, was found not to be cost-effective. Instead the base was made in PF, offering a cheap and simple replacement to acetal. The PF allowed a design which incorporated thin and thick rigid sections, to a much greater extent than any thermoplastic. The base was redesigned with thick sections to give strength and rigidity.

However, a problem arose in service with the PF base after it had performed several thousand operations. In the operation of the switch, a metal electrical contact was forced along a ramp to make contact with a second conductor, the mating contacts being preloaded on contact. Unfortunately the moving contact rubbed away a small amount of material from the surface of the ramp, which became built up around the contacts, and after long use the switch would become inoperative. This was corrected by moving the ramp away from the electrical contacts so that when the mating contact rubbed itself along the ramp, the loosened material would be away from the contact.

The replacement phenolic base was originally compression-moulded in a two-cavity mould with a cycle time of 70 seconds at a moulding temperature of $155 \pm 5°C$. After the introduction of direct screw injection feed into the compression mould, the cycle time for the component was reduced to 52 seconds.

10.2.2 Resol PF resins

Until the early 1980s, resol resins were only used in large volumes for production of paper and fabric laminates. Resols are readily soluble in water and aqueous alcohol solutions, allowing ease of impregnation of paper and many other fibrous materials. After impregnation, the material is dried at 80–100°C to remove the solvent without inducing cross-linking. Flat laminates are formed by plying up layers (or foils) of impregnated paper in a heated press. Decorative laminates (e.g. Formica™ types), include a top layer of water-white MF impregnating printed paper and a barrier layer of thin-gauge aluminium foil. The aluminium is employed to prevent strike-through by the dark-coloured PF used in the core, to promote heat transfer and minimize local distortion in the finished product, for instance if a hot saucepan is placed on a work surface. If a cigarette is allowed to smoulder, the interlayer of foil will aid heat transfer and will help reduce burn marking, PF and MF having good flame resistance. Paper laminates are widely used in the electrical industry. Thick paper and cotton fabric laminates find application as very durable gear wheels (e.g. Tufnol™ products), although a gear wheel will have to be machined from block, rather than being moulded semi-finished.

PF foams based on acid-cure resol resins and blown using physical blowing agents (n-alkanes or fluorocarbons) find use in specific applications. Originally, PF foam of low density ($20–100\,kg\,m^{-3}$) suffered from the problems of high friability and open cell structure, limiting its use as an insulation material. Development of surfactants to stabilize cell structure during blowing and cure has given rise to a closed-cell material whose insulation characteristics (based on thermal conductivity data) are as good as polyurethanes at similar densities, and appear to be retained for longer (Table 10.1). Most importantly, the PF foam provides very low flammability and smoke emission in cellular boardstock material, having no competition in fire-critical applications.

An open-cell hydrophilic grade of PF foam is used in floristry to increase the life of cut flowers. The foam will readily absorb a large volume of water, while supporting flower stems pushed into it (an example is Oasis™).

Sprayable grades of PF foam are being developed for in-situ use (for example as replacement for asbestos in building applications), curing within one minute, although reaction between the mixed ingredients (PF resol plus strong acid catalyst) can be neutralized by alkaline substrates such as concrete unless primer coats are first used. The fire properties of PF foams can be affected by choice of acid catalyst and blowing agent.

Table 10.1 Thermal conductivity of phenol formaldehyde insulation foam. By permission of Kooltherm Ltd

Age	Thermal Conductivity ($W m^{-1} K^{-1}$) Parallel*	Perpendicular*
0	0.0178	0.0176
30 days	0.0158	0.0143
19 weeks	0.0163	0.0148
11 months	0.0170	0.0155
23 months	0.0186	0.0166
35 months	0.0206	0.0181
48 months	0.0214	0.0189

(1) Testing of BS 874: 1973 (1980), Appendix B: Heat-flow Meter. Testing at a mean temperature of 10°C, storage temperature of 21°C.
(2) *Unfaced cut samples taken from parallel and perpendicular direction to original rise.
(3) PF foam based on Cellobond K™ (BP), nominal density $35 kg m^{-3}$, mixed fluorocarbons used as blowing agents.
(4) Laminated insulation board maintains a lower value of thermal conductivity for longer than the unfaced material, i.e. superior retention of fluorocarbons in the foam's cell structure.
(5) Thermal conductivity values ($W m^{-1} K^{-1}$) for other insulation materials;

Rigid PU ($35 kg m^{-3}$)	0.018 (initial and 0.023 after short-term ageing).
Expanded polystyrene ($16 kg m^{-3}$)	0.031
Glass wool ($64 kg m^{-3}$)	0.036
Expanded PVC ($40 kg m^{-3}$)	0.030 (after Brydson, ref. 4)

Intregral-skin PF foam is being produced by reaction injection moulding techniques, for specialist products such as one-piece drop-down tables for passenger aircraft.

The importance of the fire resistance, combined with low smoke emission, is now leading to the use of PF as a replacement for unsaturated polyester sheet in many glass-fibre-reinforced products. London Transport Underground service specifies PF sheet moulding compounds (PF.SMC), to the exclusion of polyester, in the manufacture of passenger seating (see also Figure 10.5). It has been necessary to develop new grades of glass fibre compatible with PF and new thickening agents (based on isocyanates for example), giving a material that will process like the polyester SMC and at relatively low temperatures. Glass-reinforced PF composites are made by hand lay-up techniques; examples of potential use include the weapons and equipment storage boxes used on British naval ships, where, following the Falklands conflict, fire retardancy and low smoke emission were found to be essential. Unfortunately, the strength of acid catalysts employed with the earlier commercial systems has caused handling and mould corrosion problems in both PF hand lay-up and PF.SMC applications. Alkali-cure PF systems seem to have overcome this problem, but have to be processed at higher temperatures.

Figure 10.5 An example of the strength combined with excellent fire resistance properties of glass reinforced PF: whole front end of the London Underground 'C' train. By permission of London Transport Ltd and BP Chemicals

With the recent introduction of PF foams and SMC materials, the application of phenolics appear to be having a renaissance, and perhaps they will regain some of their market share of 30 years ago.

10.3 Urea-formaldehyde plastics

Urea-formaldehyde (UF) and melamine-formaldehydes are known as the *aminoplasts*. Interest lies in UF, since it is based on intermediates not derived from oil, and therefore price escalation is less of a problem than for most other plastics.

Urea is obtained by reacting carbon dioxide and ammonia, the ammonium carbamate formed being dehydrated to urea (Reaction 6):

$$CO_2 + 2NH_3 \longrightarrow NH_2CO.ONH_4 \longrightarrow$$

$$NH_2CO.NH_2 + H_2O$$

Reaction 6

Urea and formaldehyde react to form mono- and dimethylol ureas, which are the usual prepolymers for UF plastics (Reaction 7):

$$O=C \overset{\textstyle N H_2}{\underset{\textstyle N H_2}{\big<}} \ + \ H.C H O \ \longrightarrow$$

$$O=C \overset{\textstyle N H.C H_2 O H}{\underset{\textstyle N H_2}{\big<}} \ + \ O=C \overset{\textstyle N H.C H_2 O H}{\underset{\textstyle N H.C H_2 O H}{\big<}}$$

Reaction 7

Cross-linking is brought about by any of a number of mechanisms (Reactions 8, 9 and 10):

$$\sim N H.C H_2.O H \quad + \quad H O.C H_2.N H \sim$$

$$\downarrow$$

$$\sim N H.C H_2 O C H_2.N H \sim \ + \ H_2 O$$

$$\downarrow$$

$$\sim N H.C H_2.N H \sim \quad + \quad H.C H O$$

Reaction 8: Methylol–methylol condensation

$$\sim N H.C H_2 \sim \ + \ H O.C H_2 \sim \ \longrightarrow$$

$$\underset{\textstyle C H_2 \sim}{\overset{\textstyle \sim N.C H_2 \sim}{|}} \ + \ H_2 O$$

Reaction 9: Methylol–imino reaction

$$\sim N H.C H_2 \sim$$
$$+ \ H.C H O \quad \longrightarrow$$
$$+ \ \sim N H.C H_2 \sim$$

$$\underset{\textstyle \sim N.C H_2 \sim}{\overset{\textstyle \sim N.C H_2 \sim}{\underset{\textstyle |}{\overset{\textstyle |}{C H_2}}}} \ + \ H_2 O$$

Reaction 10: Formaldehyde–imino condensation

UF plastics find widespread use because of their low cost and the possibility of pigmenting in a wide colour range, including pastel shades, not possible with PF materials. They have poorer water and heat resistance than PFs, only operating up to 70°C, although this can be improved by modifying with melamine. Notable properties of the UFs are the good electrical resistance, including resistance to 'tracking'—degradation of the polymer by electrical breakdown which leaves a conducting 'track' on the surface. This is important in electrical plugs and sockets. Foodstuffs are not tainted by UF, and so it can be used for caps and closures; good resistance to glycol and alcohol also promotes the plastic's use in these applications. UF is favoured in the use for caps and closures, especially for its 'feel' of the component and its visual appeal. Technically UF provides products which are strong and rigid, and reasonably immune to attack by a wide variety of fluids, which in turn will not be contaminated by the plastic. Figure 10.6 shows a selection of metallized UF closures.

Wood-flour is normally used as a filler, except for very light colours where the purer α-cellulose is employed. The bulk of UF mouldings are found in caps and closures of bottles and other containers, and in electrical fittings. Substantial quantities are still used in applications as diverse as buttons and toilet seats. UF resins are used as wood adhesives in the manufacture of plywood and chipboard. Acid-catalysed UF foams (U-foam) are used to fill cavity walls on site, competing successfully against PUs, PFs and mineral wools, UF foam is a highly-closed cell system at densities less than $10 \, \text{kg} \, \text{m}^{-3}$,

Figure 10.6 Caps and closures moulded by Metal Closures Mouldings Ltd from urea-formaldehyde moulding powder supplied by BP Chemicals Division and subsequently metallized

but because it is very friable, use is limited, needing the protection of the double layer of cavity brick wall to prevent rapid mechanical breakdown. Foams are also used as firelighters.

10.4 Melamine-formaldehyde plastics

Melamine is obtained from calcium cyanamide via dicyandiamide by a sequence of reactions (Reaction 11);

$$Ca(CN)_2 \longrightarrow NH_2.CN \longrightarrow NH_2.C.NH.CN$$

with the middle/right structure showing the group:

$$\underset{\underset{NH}{\overset{\|}{}}}{NH_2.C.NH.CN}$$

Reaction 11

Melamine and formaldehyde react to yield methylol-melamines, the degree of methylolation depending on the melamine/formaldehyde ratio. The principal cross-linking reaction involves methylol-methylol condensation to —CH$_2$—O—CH$_2$—, although methylene links may be formed by the elimination of formaldehyde from the bridge or by methylol-amine condensation, the first two reactions being similar to those of UF prepolymer's cross-linking mechanisms.

Melamine-formaldehyde (MF) mouldings compare favourably with their UF counterparts in all properties. The plastic has better heat resistance, lower water absorption, better stain resistance and higher hardness and abrasion resistance, but its price is higher.

An important application is in tableware (Figure 10.7), picnicware and, more recently, kitchenware (Figure 10.8), in all of which tactile and aesthetic properties are important along with good mechanical and chemical properties. MF is used to impregnate the surface lamination of decorative laminates, to give an attractive product of very high hardness, such as Formica™ and Warerite™ types. Glass-reinforced MF laminates are used in good-quality electric boards, having heat resistance to 200°C and good electrical properties, including tracking resistance.

Figure 10.7 Melamine moulding powders are used for this range of tableware

Figure 10.8 Melamine moulding powders are used by Addis for its range of kitchenware

10.5 Unsaturated polyester resins

There are many different types of hydroxylated and dicarboxylic acid monomers that will react and form ester linkages. In section 7.3, the saturated thermoplastic polyesters such as poly(ethylene terephthalate) were considered, and materials such as poly(propylene adipate) are used as plasticizers for PVC. It should be noted that polymers with the ester grouping in their side chains, such as poly(vinyl acetate) or poly(methyl methacrylate), are not classified as polyesters. Variation of the starting monomers permits a limited choice of properties to be built into the structure of the final polymer.

An unsaturated dicarboxylic acid provides two types of functionality. It can be partially polymerized by a step-growth condensation reaction with a glycol, (in non-stoichiometric proportions) to yield a polyester oligomer or prepolymer with residual unsaturation in its chain. The unsaturation can be reacted at some later stage, by free radical initiators permitting a limited chain reaction with a low-molecular-weight unsaturated monomer (styrene, for instance, being a common *cross-linking agent*). There are usually three constituents in the prepolymer:

(i) *Glycol*, frequently propylene glycol (propane-1,2-diol) to prevent the polyester crystallizing by steric hindrance (cf. ethylene glycol)
(ii) *Saturated acid* to reduce the degree of cross-linking, usually either one of the phthalic isomers, or adipic or sebacic acids if flexibility is required
(iii) *Unsaturated acid*, which is always maleic acid (or maleic anhydride) or its *trans* isomer, fumaric acid.

Normally the unsaturated acid/saturated acid ratio is between 1:1 and 2:1.

Polyesterification is appropriate to the acid mixture employed and is carried out to a molecular weight of about a thousand. The resulting resin is extremely viscous, if not solid, at room temperature and must be diluted with a liquid vinyl monomer, either styrene or a styrene/methyl methacrylate mixture, which will subsequently participate in the cross-linking reaction. For isophthalic and terephthalic acid-based systems, polyesterification is usually carried out in two stages. First, the saturated phthalic acid is reacted with glycol to produce a very low-molecular-weight polyester chain with hydroxyl termination (called *tipping*). In the second stage, this product is treated with unsaturated acid (or anhydride) to yield the unsaturated polyster oligomer; this is dissolved in styrene. The mixture is usually a pale amber, syrupy liquid and contains inhibitors to prevent premature cross-linking. Quality control tests include density, viscosity, acid number and colour.

The cross-linking reaction is initiated by peroxide catalyst at 50–150°C, or by a redox system at room temperature. Initiators include benzoyl peroxide and *t*-butyl perbenzoate for high-temperature curing, or methylethylketone peroxide and cyclohexanone peroxide, with cobalt octoate or naphthenate as accelerator for low-temperature processes. Photo-initiators are used in thin film applications, such as UV curable surface coatings.

10.5.1 Structure and properties

The cross-linked unsaturated polyester resin (UPR) is a rigid product, but its detailed microstructure is less well defined. If maleic acid is used in the condensation reaction, most of it isomerizes to the more favoured *trans*-isomer (fumarate), which reacts differently with styrene, as shown by the different copolymerization reactivity ratios in 'model' reactions (Table 10.2). The data show that in the styrene/maleate system, the styrene is more likely to homopolymerize; this is clearly undesirable when employed for cross-linking. The styrene/fumarate reaction is more favoured for copolymerization and hence for cross-linking.

The uncertainty of the maleate/fumarate ratio in any polyester oligomer, may account for reactivity differences in UPR, since reaction rates are also different. Cross-linking density can be adjusted by varying the unsaturated/saturated acid ratio, and the rigidity can be increased by going from an aliphatic to aromatic acid. Ester groups provide sites for hydrolytic degradation, particularly in alkaline conditions, limiting the use of simple resins in corrosive environments. This has lead to the development of a wide variety of cross-linked plastics with superior hydrolytic stability. This property can be improved by replacing *ortho*-phthalic acid by the *meta*- or *para*-isomers, isophthalic and terephthalic acid respectively. Further improvements can be obtained by more extreme changes in the chemistry, such as in the vinyl ester resins, although these products are considerably more expensive than the conventional UPR. The structure of the polyester, especially after reacting with styrene, promotes poor fire resistance. External fire retardants, such as halogenated phosphorus compounds or antimony trioxide have to be used in the event of fire risk; alternatively, chlorinated dicarboxylic acids can be used in the synthesis of the unsaturated polyester chain.

10.5.2 Glass fibre-reinforced polyesters (GRP)

UPRs are normally rigid, brittle materials and as such cannot be used alone to form moulded products. Instead the polyester has to be reinforced with fibrous material having a tensile modulus at least ten times that of the cured resin: the resin may be considered as an adhesive matrix binding the fibres together.

Table 10.2 Reactivity ratios of styrene with unsaturated ester groups

	Styrene/diethyl maleate	Styrene/diethyl fumarate
r_1	6.52	0.3
r_2	0.005	0.07

Glass fibre is a preferred form of reinforcement, providing strong laminates at an economical price. Several forms are important. Glass cloth is expensive and will only be used for the most critical applications: various weaves are available. Chopped strand mat consists of bundles of glass filaments, approximately 50 mm long, bonded by a resinous binder. Other forms of glass fibre including needle mat, rovings, yarns, tapes, surfacing mats and continuous filaments are available. Short-length hammer-milled and chopped strand fibres (0.2 to 5 mm) are also used in special products. The glass may be of two types; *E glass*, a low-alkali borosilicate for electrical applications, and *A glass*, a cheaper variety for general use, with an alkali content of 10–15%.

It is usually necessary to provide a finish for the fibres to facilitate good wetting and interfacial bonding between the glass and the resin. Most important is vinyl trichlorsilane treatment, which gives direct chemical linkage between glass and resin:

Binder materials for chopped strand mats include starch, poly(vinyl acetate) and polyesters. It is important that the binder is compatible with the resin and the end-use of the laminate. In the production of glass/resin composites, several rules must be obeyed to achieve optimum mechanical properties. These include maximum glass-to-resin ratio and the establishment of good wetting and air release in the composite's structure.

The chief process for manufacturing large structures is the hand lay-up technique. The method allows large structures to be made without complicated moulds; the resulting product has superior heat resistance compared to many thermoplastics, and the strength and rigidity are good (2–10 times better than most thermoplastics), at densities less than that of most metals (1.4–2.2 Mg m^{-3}).

There have been many attempts to reduce the very high labour content of the hand lay-up technique, and similar efforts to reduce the cost of the materials involved. An important attribute of the process is that the reinforcing fibres can be used very effectively by disposing them geometrically to counter the greatest stresses in use. This is unlike many other moulding processes where there is a risk of undesirable fibre orientation. It should be noted that maximum interlaminar strength is achieved by laying-up subsequent laminates while the substrate is in a near gel state but still has a tacky and therefore active surface. Ideally, there should be limited interpenetration of fibre across the interface without any major disturbance of the rest of the glass fibre.

Figure 10.9 *Arun* class lifeboat in GRP under construction

Figure 10.10 Hand lay-up of polyester GRP, used in the production of panels for exterior kiosks. By permission of WES Plastics Ltd

Examples of the strength, durability and potential size of polyester laminates can be found in the marine field, e.g. in the construction of boats and ships: notable examples are the 'Arun' class lifeboat in service with the Royal National Lifeboat Institution (Figure 10.9), and minesweepers in service with the Royal Navy. Relative sizes are 17 and 60 m, the latter using 300 tonnes of UPR.

There is controversy as to whether spray-up methods give GRP laminates as good as those obtained by hand lamination, but there is little doubt that the products are adequate for many applications, including leisure craft. Figure 10.10 shows one stage in the production of panels used in the fabrication of semi-permanent exterior *kiosks*. Gel coat can be spray-applied to the released surface of the mould in a small fraction of the time taken by hand. This is then followed by a mixture of glass fibres and catalysed resin, the former being obtained by chopping 'continuous glass rovings' fed to the spray head. Proponents of the technique claim that there are economies in time (cutting of the chopped-strand mat to shape is eliminated and spray application is claimed to be faster than brush and roller impregnation), and in cost (the continuous rovings being cheaper than chopped-strand mat).

As an alternative to the 'bucket and brush' technique of resin impregnation or spray application and their associated limitations in production control, resin injection has been developed. A matched mould system capable of withstanding high pressures has to be used, and so limits product size, the largest commercial mouldings being produced are car bodies made in limited numbers (e.g. Lotus cars). Another problem is that the viscosity of the resin must be low and there is the operating problem of ensuring that the catalysed resin does not gel in the injection unit. The advantages of resin injection includes dimensional repeatability, maximizing glass/resin loading (60–80% glass in the composite), air removal and the styrene liberated into the atmosphere can be very much reduced and therefore contained.

For the methods of application above, resin viscosity must be maintained low, and this precludes the use of particulate fillers as inert diluents to cheapen the system. Sheet-moulding compound (SMC) and dough-moulding compound (DMC) prepolymer forms of UPR continue to be a major growth area of GRP technology.

SMC is prepared by blending uncured UPR resin, with styrene, heat-activated catalyst, inert particulate fillers, internal release agent and low profile aids. In the final stage of mixing, MgO or $Mg(OH)_2$ is blended into the paste. This compound will gradually react with the polyester, either by hydrogen bonding or more probably by additional esterification. As a result, the viscosity of the paste will increase within hours from 5 to 500 Pa s and the material will be handleable, being essentially dry to the touch. The combination of the catalyst and the MgO *thickener* gives the UPR a short shelf-life, normally less than 3 months. However, before the viscosity increases the resin blend must be coated on to polyethylene or polyamide film (to facilitate

Table 10.3 Typical polyester SMC and DMC recipes

	Weight (%)	
	SMC	DMC
Polyester resin	30.0	25.0
Peroxide initiator	0.5	0.5
Calcium carbonate	33.5	45.5
Low shrink/profile additive	6.0	5.0
Release agent (e.g. zinc stearate)	2.0	2.0
Pigment	2.0	2.0
Thickener (e.g. MgO)	1.0	—
Chopped glass fibre	25.0	20.0

handling), chopped glass fibre (50 mm long) is randomly deposited on to the resin and a second resin-coated film applied on top. The sandwich is compacted between rollers to speed wetting of the glass and to give uniform thickness to the sheet. The sheet is rolled up and allowed to *mature* at 40°C for 48 h. The final glass fibre loading will be 25–40% of the SMC. Typical formulations of SMC and DMC are given in Table 10.3.

Solid thermoplastic polyester (melting at 80–90°C), has been used as an alternative thickener to MgO. This material is claimed to increase the shelf-life of the SMC considerably, but its 'melt viscosity' in processing is much lower than the norm, leading to an increased risk of glass fibre/resin separation in longer flow paths.

In moulding, the film is removed and the SMC cut for charging into the mould: several plies can be built up for thicker moulding sections. The SMC is shaped by press tools, heating causing initial flow and subsequent cross-linking. The viscosity of the SMC is sufficiently high for the glass fibres to be carried with the resin in the mould, but care has to be taken to prevent excessive flow which will cause orientation in the fibres and in some instances fibre dilution, producing weaknesses in the product. A further consideration is that the charge shape should be obtainable without waste from the SMC sheet available. Currently SMC use is limited by;

(i) Limited thickness of uncured stock, 10 mm being the maximum, although this is not considered a real problem by vehicle manufacturers, for example

(ii) The ability to predict flow patterns during shaping, particularly to reduce weld-line failures

(iii) Design of blanks to give optimum properties

(iv) Surface defects, resulting from air traps blistering in post-moulding operations. Low profile additives and in-mould coating techniques have provided only limited relief from the defects. Usually costly and time consuming post-moulding filling has to be carried out before a cured SMC component can be painted.

Although, on mechanical property grounds, it is desirable to have restricted flow (i.e. it is normal to have no more than 50% increase on the original charge area), the surface finish of the moulding is improved by encouraging more flow, that is by using a SMC charge of lower area and greater thickness. There is some development of SMC as a lightweight load-bearing material. Polyester SMC is used largely in the manufacture of semi-structural vehicle components, particularly for body panels for truck cabs, car bonnets, roofs and fenders (see Figure 11.9). Apart from automotive use, SMC finds application in building (window frames, meter boxes); furniture (outside tables, chairs); and electrical components (switch boxes, electrical cabinets, road-lamp housings, television cabinets). Other large SMC units moulded in a single piece include satellite receiving dishes, which are about 1 m in diameter (Figure 10.11). SMC is normally transparent to many frequencies of radio-waves, and so has to be surface coated with a special reflective paint to direct satellite signals to the dish's receiver.

Dough moulding compounds (DMC) are prepared by blending resin, filler, glass fibre (10–25 mm long), pigment and internal release agent in a dough mixer; the resulting material has the consistency and handling characteristics

Figure 10.11 Domestic satellite receiving dish, based on polyester SMC. By permission of BTR Permali Ltd

Figure 10.12 British Rail passenger seat frames moulded in DMC.Photograph: British Rail

of a bread dough. DMC can be compression- or injection-moulded, although pressures required are low, 15–75 bar. Polyester DMC is popular for the manufacture of mechanically strong electrical insulants, such as domestic plug tops and sockets, now available in a variety of bright colours. It is also used to mould large and strong articles such as passenger seat frames, for British Rail (Figure 10.12).

The overall pattern of use of polyester GRP is given in Table 10.4. Competition in the automotive industry is coming from phenolic and thermoplastic SMC, and certain engineering thermoplastics.

10.6 Epoxide resins

In comparison to the polyester resins, epoxides have limited and specialist use, and cost at least twice as much. However, they have superior toughness, will operate at slightly higher temperatures, have excellent adhesion to many substrates and are particularly resistant to alkaline environments. In particular, the chain extension and cross-linking reactions do not normally

Table 10.4 Uses of polyester GRP

	Weight (%)
Building and construction	15
Land transport	26
Marine	22
Tanks, pipes, corrosion-resistant applications	10
Electrical	8
Other industrial	8
Other applications (e.g. consumer goods)	10
Aircraft	1

produce side products, and there is less shrinkage compared with other thermosetting plastics. This allows mouldings of superior quality and dimensional tolerance to be manufactured, especially in the gate area.

The most important epoxide resins are oligomers produced from the reaction of bisphenol A and epichlorhydrin. Bisphenol A is prepared from phenol and acetone, and need not be pure. Epichlorhydrin is made from propylene via the hydrochlorination reaction with propylene oxide. *Diglycidyl ether* is produced in the reaction of bisphenol A with excess epichlorhydrin (2–3 times stoichiometry), in an alkaline solution to prevent higher homologues being formed ($n = 1$. in Reaction 12).

Reaction 12

Chain extension and cross-linking mechanisms involve attack on the epoxy groups. Higher homologues ($n > 1$) are favoured by more alkaline conditions and reduced excess of epichlorhydrin (Reaction 13);

$$\sim\overset{O}{\overset{\diagup\diagdown}{C}}H.\overset{}{C}H_2 \ + \ HO\sim \overset{NaOH}{\longrightarrow} \sim\overset{O}{\underset{|}{C}}\overset{H}{\underset{}{C}}H.CH_2.O\sim$$

Reaction 13

These reactions favour the formation of hydroxyl group along the chain, broadening the choice of curing agent to produce cross-linking. In addition, many analogous resins have been derived, typically from epichlorhydrin and a variety of compounds having active *hydrogen* atoms – usually hydroxyl compounds. Epoxide resins are characterized by viscosity (for liquid resins, $n = 1$–3), epoxide equivalent, hydroxyl equivalent, molecular weight and distribution, melting point of solid resin and heat distortion temperature of the cured resin.

Cross-linking can proceed through a number of mechanisms. Tertiary amines and boron fluoride complexes catalyse homopolymerization (Reaction 14);

$$R_3N \ + \ \overset{O}{\overset{\diagup\diagdown}{C}}H_2.CH\sim \ \longrightarrow \ R_3\overset{+}{N}.CH_2.\overset{O^-}{\underset{}{C}}H\sim$$

$$\downarrow \quad \overset{O}{\overset{\diagup\diagdown}{C}}H_2.CH\sim$$

$$R_3\overset{+}{N}.CH_2.\overset{H}{\underset{O}{C}}\sim$$
$$\overset{\diagdown}{C}H_2.\overset{C}{\underset{O^-}{C}}H\sim \ \longrightarrow \ etc.$$

Reaction 14

Primary and secondary amines are employed (as Reaction 15);

$$\sim\overset{O}{\overset{\diagup\diagdown}{C}}H.CH_2 \ + \ \overset{R}{\underset{H \diagup \ \diagdown H}{N}} \ + \ \overset{O}{\overset{\diagup\diagdown}{C}}H_2.CH\sim$$

$$\downarrow$$

$$\sim\overset{OH}{\underset{|}{C}}H.CH_2.\overset{R}{\underset{|}{N}}.CH_2.\overset{OH}{\underset{|}{C}}H\sim$$

Reaction 15

Acid and acid anhydride hardeners are less volatile and so safer to use. They give lower exotherms, but will react both with the epoxy and hydroxyl groups in the oligomers (Reaction 16);

$$\sim\!\overset{O}{\underset{}{C}}H.\overset{}{C}H_2 \;+\; H\,O\,\overset{O}{\underset{\|}{C}}.\,R\,.\overset{O}{\underset{\|}{C}}\,O\,H \;+\; \overset{O}{\underset{}{C}}H_2.C\,H\!\sim$$

$$\sim\!\overset{OH}{\underset{|}{C}}H.C\,H_2.O\,\overset{O}{\underset{\|}{C}}.R.\overset{O}{\underset{\|}{C}}\,O.C\,H_2.\overset{OH}{\underset{|}{C}}\,H\!\sim$$

Reaction 16 (acids can also react with hydroxyls)

The availability of hydroxyl groups in the epoxide molecule permits reaction with isocyanates and PF oligomers, particularly in the manufacture of surface coatings. Many of the reactions shown above are slow, needing high temperatures to produce complete cure. Applications include mouldings where good properties and tolerances are essential, for example the encapsulation of electronic components; however, epoxides are meeting some competition from cheaper polyurethane resins. Metal shaping moulds, prototype low-pressure moulds for epoxides, polyurethanes, thermoforming moulds, etc. are often based on reinforced epoxide resins with high loadings of cheap filler to maximize life and reduce cost. Resin-based tools are considerably cheaper to fabricate than steel and aluminium alloys, lending themselves to be built-up or cut away, if small changes are required. The largest uses of epoxides are in surface coatings and adhesives, particularly for metal–metal and metal–plastics bonding. Floor coatings, aggregate-filled road surfacing and marine coatings (for example for oil rigs), with heavy wear resistance, are particularly important uses of epoxides as coatings. Epoxide resins are used in fibre composites where polyester or silicone resin composites will not suffice, for example where environmental problems, fatigue life and creep resistance are critical. Examples include carbon-fibre-reinforced leaf-springs for public transport. For ease of handling, a thixotropic resin (pre-mixed with heat-activated curing agent) is used to pre-impregnate glass cloth, which can be cut and shaped in a heated mould, similar to SMC. The *pre-pregs* eliminate the handling problems of liquid resins, especially for those moulders who need only small quantities of material, such as in passenger aircraft flooring.

References

1. Brydson, J.A., *Plastics Materials*, Iliffe [Butterworth], London (1982).

11 Polyurethane plastics

11.1 Introduction

The polyurethanes (PU) offer the widest choice of properties of any group of plastics materials, and the opportunity to incorporate specific chemical structures into a PU chain allows the polymer chemist to tailor many specific properties into the polymer. Therefore thermoplastic, thermoset resinous and elastomeric PUs are available, many of them in cellular forms. Specific PU properties include abrasion resistance and general ease of processing.

The choice of properties stems from the vast number of monomeric and oligomeric ingredients that can be reacted together to produce polymers containing the urethane group in their molecular structure. The urethane linkage results from the reaction between isocyanate and hydroxyl-containing compounds; catalysts and/or heat accelerate the reaction:

$$\sim R'.N{=}C{=}O \;+\; \sim R'.O{-}H \longrightarrow$$

$$\sim R'{-}\underset{\underset{H}{|}}{N}{-}\underset{\underset{O}{\|}}{C}{-}O{-}R''\sim$$

The reactivity of many isocyanates is exploited by promoting side reactions, either within the polymer chain or at chain ends, with reactive additives such as chain extenders or cross-linking agents. The ease with which many monomeric ingredients are converted into polymer during the shaping operation permits relatively simple and cheap processes to be employed, often with low expenditure of energy.

PUs find use virtually in any field of application: products can be shaped by traditional thermoplastic processes, and by coating or calendering on to substrates. However, the major manufacturing techniques involve *dual component processing* of the liquid 'monomeric' ingredients. Products made by such methods include slabstock and moulded foams; self-skinned foamed elastomers and resins, and non-foamed mouldings made by reaction injection moulding; surface coatings, sealants, encapsulants and adhesives. Foamed PUs are used in a wide range of products from cushioning materials for furniture and automotive seating, to thermal insulation, structural stiffening for building panels or refrigerators. High-quality mouldings include automotive panels (glass-fibre reinforced), head restraints, steering wheels, car trim and covers for electronic equipment. References listed at the end of this chapter give comprehensive details of the properties of PU materials.

11.2 Molecular structure–property relationships

A polyisocyanate and a polyol (a polyhydroxyl material) are the basic ingredients that react to form a polyurethane: polyols with functionalities from 2 to 6 are commonly used. Low-molecular-weight polyols (functionality 2 and higher) or chain-extending diols, will tend to impart rigidity into a PU. High-molecular-weight polyols, (functionality 3 or less) will tend to give softer, low-modulus PUs. It follows that the ratio of polyol to isocyanate (the stoichiometry that will allow a reaction to form a full urethane polymer) is high for high-molecular-weight polyols and low for low-molecular-weight polyols. In addition, the amount of isocyanate reacted with polyol can be varied around the theoretical stoichiometry, that is, by changing the *isocyanate index*. A low *isocyanate index* may tend to encourage plasticization of the PU because of the presence of excess polyol, and a high *isocyanate index* will promote stiffening through cross-linking side reactions (see section 11.2.1). However, to produce thermoplastic PUs, reactants with a functionality of 2 mixed at precise stoichiometric proportions must be used. Cellular PUs are relatively cheap to manufacture, and dominate the marketplace, even although the cost of raw materials may be at least twice that for foamed commodity thermoplastics.

11.2.1 *Typical reactions*

Isocyanate under suitable conditions will readily react with active hydrogen-containing compounds.

$$\sim R'.N{=}C{=}O \quad + \quad H_2O \quad \longrightarrow \quad \sim R'.NH_2 \; + \; CO_2 \nearrow$$

(i) Isocyanate/water

$$\sim R'.N{=}C{=}O \; + \; \sim R''.NH_2 \quad \longrightarrow$$

$$\sim R'{-}\overset{\displaystyle H}{\underset{}{N}}{-}\overset{\displaystyle O}{\overset{\|}{C}}{-}\overset{\displaystyle H}{\underset{}{N}}{-}R''\sim$$

(ii) Isocyanate/amine (polyurea formation)

$$\sim R'.N=C=O \quad + \quad \sim R''-\overset{\overset{\displaystyle H}{|}}{N}-\overset{\overset{\displaystyle}{\underset{\underset{\displaystyle O}{||}}{C}}}-\overset{\overset{\displaystyle H}{|}}{N}-R'''\sim \quad \longrightarrow$$

$$\sim R''-\overset{\overset{\displaystyle O}{||}}{C}-\overset{\overset{\displaystyle H}{|}}{N}-R'''\sim$$

(iii) Isocyanate/urea (polybiuret formation)

$$\sim R'.N=C=O \quad + \quad \sim R''-\overset{\overset{\displaystyle}{\underset{\underset{\displaystyle H}{|}}{N}}}-\overset{\overset{\overset{\displaystyle O}{||}}{C}}{}-O-R'''\sim \quad \longrightarrow$$

$$\sim R''-\overset{\overset{\displaystyle}{\underset{\underset{\displaystyle C}{|}}{N}}}-\overset{\overset{\overset{\displaystyle O}{||}}{C}}{}-O-R'''\sim$$

(iv) Isocyanate/urethane (polyallophanate formation)

$$3 \ \ O=C=N-R-N=C=O \quad \longrightarrow$$

(v) Isocyanate/isocyanate (e.g. isocyanurate formation)

The side reactions (iii-v) lead to cross-linking, and will readily occur with the polyol/aromatic isocyanate reactants at ambient conditions, if both the appropriate catalyst and slight excess isocyanate are used, even though the isocyanate may be di-functional. Urethane cross-links also readily occur with polyols or isocyanates of functionality 3 or higher.

11.2.2 Monomeric components

(i) *Polyols.* Typically these are *oligomers* derived from several classes of hydroxyl-containing material. Polyether polyols are typically made by reacting ethylene or propylene oxides with water in the presence of alkaline catalysts; these are made in the molecular weight range of 600 to 6 000.

$$n \; H_2C \overset{O}{\overbrace{}} C.H.R \quad + \quad n \; H_2O \longrightarrow$$

$$H \; O \text{-} \!\!\left[C H_2 . \overset{R}{\underset{}{C}}.H \text{-} O \right]_n \!\! H$$

$$(R = \; H \; , \; C H_3)$$

Polyethers are used in the manufacture of flexible foams, giving good resilience and hydrolytic stability, and are the basis of many PU.RIM systems, although they have limited thermal stability (to a maximum of 70°C). Polyester polyols are made by conventional polyester processes (for example adipic acid reacted with butane-1, 4-diol), although the diol is used in slight excess to ensure that the resulting oligomer has hydroxyl-tipped chain ends: again, the molecular weight range is between 600 and 6 000. Speciality polyesters are made from the reaction of ε-caprolactone with low-molecular-weight diols (e.g. ethylene glycol) or triols (e.g. glycerol).

Polyester polyols impart better mechanical properties and thermal stability to the PU, but have poorer hydrolytic resistance than the polyethers. Polyesters are used in large tonnages for shoe soling, some flexible foams, and many elastomer formulations. Polyols with functionality greater than two are produced from a wide number of sources, such as propylene oxide reacted with glycerol or sugars (for polyols of functionality 3 and 6, respectively), and other naturally-occurring materials. Sugar-based polyols are used to give high rigidity in low-density structural foams, although higher-molecular-weight polyether polyols may be included in a polyol blend to reduce brittleness. The ease with which many other compounds can be further hydroxylated broadens the variety of starting monomers; polybutadiene, for example, is polyhydroxylated to permit isocyanate vulcanization. Polyethers are largely low-viscosity liquids at 15°C, permitting easy processing; polyesters tend to be either viscous liquids or waxy solids and need to be heated to effect processing. Polyols are characterized by hydroxyl value, water content and molecular weight.

(ii) *Isocyanates.* The choice of polyisocyanates is more limited. Of the aromatic isocyanates, toluene diisocyanate (TDI), and 4, 4'-diphenylmethane diisocyanate (MDI) are commonly used.

$$CH_3$$

N=C=O   TDI

N=C=O

$$O=C=N-\bigcirc-\overset{\overset{H}{|}}{\underset{\underset{H}{|}}{C}}-\bigcirc-N=C=O \quad \text{MDI}$$

TDI is made by the nitration of toluene, to give dinitro derivatives, which are hydrogenated to the diamines and these in turn are phosgenated to the diisocyanates. Typically, a 80:20 mixture of the 2, 4 and 2, 6 isomers of various purities is employed; they are low-viscosity liquids down to 0°C. MDI is made by the phosgenation of diamino-diphenylmethane, itself formed by the condensation of aniline and formaldehyde. Pure MDI is an unstable solid (melting point 42°C), readily forming dimers and trimers. 'Polymeric' or 'crude' MDIs are fractions distilled from the reaction products, having functionalities of 2 to 3. The so-called commercial 'pure' MDIs are eutectic mixtures of pure MDI and products derived from the reaction of excess pure MDI with diol (e.g. butane-1, 4-diol); these have a melting point of 10–20°C and a functionality of 2. Aromatic isocyanates are relatively expensive (£1000 to £1500 per tonne). Isocyanates are characterized by isocyanate content and chloride content.

Unfortunately, the aromatic isocyanate component in a PU chain may contribute to poor oxidative resistance, particularly under UV-radiation. Employing mixtures of stabilizers has limited success in eliminating embrittlement or yellowing. Alicyclic and aliphatic isocyanates give good resistance to UV, (e.g. isophorone diisocyanate), finding use in products for exterior applications (e.g. surface coatings) but are more than five times as expensive and tend to be less reactive than the aromatic types.

The high reactivity of isocyanates together with their volatility gives rise to health and safety problems. A solution to volatility is to produce isocyanate-tipped PU *pre-polymers*, by reacting excess isocyanate with polyol under controlled conditions: the oligomer can react with active hydrogen species some time later. This will also permit controlled development of the structure of a PU, especially when using mixtures of polyols.

(iii) *Other additives.* *Chain extenders* are normally low-molecular-weight diols which impart limited rigidity into the PU chain. Di- and triethylene glycols, mono- and di-propylene glycols and butane-1, 4-diol are most commonly used. *Cross-linking agents* include high functional polyols, and diamines (functionality 4). Polyurea–polyurethane copolymers with superior properties are produced from 3, 3'-dichloro-4, 4'-diaminodiphenylmethane (MOCA) although it is a suspected carcinogen. Polyamines derived from propylene oxide are used in the high-speed polyurea/polyurethane reaction injection moulding process in the manufacture of automotive panels (see section 11.3.3).

Most PU processes involve the simultaneous making and shaping of the polymer at room temperature and employ *catalysts*. The urethane reaction is catalysed by organic metal compounds, typically tin salts and tertiary amines. The water/isocyanate reaction is catalysed by amines; again the tertiary amines are preferred over primary or secondary amines, since they have little odour and do not react with isocyanate. Many side reactions are promoted by heat (e.g. from reaction exotherm) and by catalysts, for instance in the formation of the allophanate cross-link. Residual catalyst in PU products will accelerate degradation in adverse conditions.

The water–isocyanate reaction forming carbon dioxide is commonly used to form open-cell flexible foams. Trichlorofluoromethane (TCFM) is a *physical blowing agent* relying on heat from the urethane reaction exotherm to boil: this is used to form rigid closed-cell foams and most self-skinned moulded products. There is considerable concern over the effect of fluorocarbons on the Earth's ozone layer, and suitable replacements are actively being sought.

Surfactants acting as nucleating agents are used to control foam cell size and distribution; materials including modified silicone oils are dissolved into the polyol ingredient stream. Rigid foamed PUs which have to be painted following moulding will include dissolved air-in-polyol as a nucleating agent, to ensure subsequent paint adhesion. *Pigments* are used to give solid colour to products, although surface coatings for the moulded product are now used, because MDIs are dark and contribute to poor UV stability of the PUs. *Fillers* are used in reaction injection moulded (RIM) products reinforced with glass fibre, used to reduce high coefficient of thermal expansion and creep (Table 11.1). The glass is in the form of short hammer-milled or chopped-strand fibre (0.2–3.0 mm in length), long enough to give reinforcement (aspect ratio greater than 10), but short enough to be pumped through dispensing

Table 11.1 The effect of glass fibre on the properties of high-modulus polyurethane elastomer[1]

Property	Orientation[2]	Glass content (%wt)			
		0	10	20	30
Density (kg m^{-3})		1100	1150	1200	1300
Flexural modulus	Par	350	600	1000	1500
@ 23°C (MPa)	RA	—	400	550	700
Coeff. of thermal	Par	180	70	40	25
expansion ($\times 10^{-6}$/°C)	RA	—	165	140	130
Heat sag, 1 h at 160°C	Par	15	8	5	3
(mm)	RA	—	11	10	8
Elongation at break	Par	320	250	180	5
(%)	RA	—	250	200	140

Notes:
1. Based on information for glass-reinforced Bayflex™ GR 110/50 (using 0.2 mm length hammer-milled glass fibre); Bayer leaflet PU52113e. For further details, see Seel, K. and Klier, L., Reinforced polyurethanes for car body parts, *Kunststoffe* 71 (1981) 9, 577–584.
2. Orientation based on the flow direction from injection point of the mould: Par, parallel to flow; RA, at right-angles to flow.

machinery without attrition of fibres or abrasive wear of the equipment. The glass fibre used in reinforced flexible PUs does not have to be surface-treated, since some fibre slippage is desirable in the flexible PU matrix.

11.3. Properties and applications of polyurethanes

11.3.1 *Flexible polyurethane foams*

The major use of PUs is in flexible foams, for upholstery, mattresses and vehicle seat cushioning. Semi-flexible grades are used as in-fill materials, for instance as an energy-absorbing composite backing plasticized PVC skins on car crashpads (part of the dashboard).

Cushion foams are made by both slabstock and moulding techniques. In large-volume slabstock production, the *one-shot* process is used; polyol, isocyanate, catalysts, blowing agents (such as water) and surfactants are metered separately and pumped to a mechanical mix-head, where the ingredients are mixed rapidly together and poured either into a moving 'trough' (Figure 11.1), or a large box mould, making slabstock *loaves* or *buns* respectively. The ingredients react and expand to fill the trough or box; the foam will be fully expanded and the surface of the foam dry to the touch in less than 10 minutes, although complete cure will take longer. The freshly-made blocks are stored in warm rooms to speed complete cure and allow volatiles to evaporate. The blocks are cut by bandsaw to the required shapes; waste PU foam can be reconstituted by gluing foam crumble together, and this can be used as a stiff infill for seating. The cut foam can then be covered with fabric to make seating, or mattresses. The inherently poor fire retardancy of the PU can be improved by incorporating halogenated phosphorus compounds, or by covering with fire-retardant fabric or interliners.

Moulded cushions tend to have slightly better fire retardancy and better resilience properties than slab-stock products. Hot-cure and, more recently, cold-cure moulding techniques have allowed better control of the side reactions (such as formation of isocyanurate and allophanate cross-links). One-shot processing with low-pressure dual-component dispensing equip-

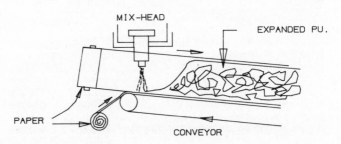

Figure 11.1 Diagram of continuous foam slab stock production

ment is used, the polyol being pre-blended with all the other ingredients except the isocyanate. For better control of the reaction and residual by-products like unreacted isocyanate, isocyanate-tipped prepolymers are used, in the so-called *two-shot* process; the prepolymer is reacted with water and polyol in the foaming reaction. The mixed ingredients are poured from the mix-head into a heated mould coated with release agent, the lid of which is closed as the reaction starts. Heating will ensure, and accelerate, important side-reactions (Figure 11.2). Demould is within ten minutes. Changing the stoichiometry of the ingredients by closed-loop control of pumps allows the manufacture of products such as up-market car seats with foamed sections of different moduli to be made in a single moulding operation. The density of the open-cell PU foams ranges from $20-100\,kg\,m^{-3}$. Until the late 1970s, TDI was the

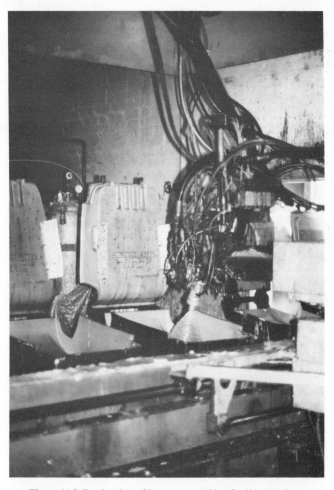

Figure 11.2 Production of hot cure moulded flexible PU foam

dominant isocyanate used, but as a result of increasing concern over the health risks, MDI and TDI/MDI blends are now more popular.

11.3.2 Rigid polyurethane and polyisocyanurate foams

Rigid PU resins can be readily foamed to low densities, giving excellent insulation and imparting stiffness to many products such as panel walls and refrigerators. Again, the problem of poor fire resistance has led to the use of halogenated phosphorus compounds as fire-retardants and also to the development of PU/polyisocyanurate foams. Slabstock techniques are used, where the cut closed-cell foam can later be adhesively laminated with any of a number of facing materials such as plasterboard, paper, or metal foils. A popular manufacturing technique involves pouring the mixed liquid ingredients directly on to a continous band of facing material, allowing partial expansion of the foam before applying an upper facing; full expansion and cure of the PU will then be accelerated by heat (Figure 11.3). Rigid foam can be injected easily into any cavity formed in a product: between the two skins of a domestic refrigerator, or into the void formed by the two sheet-steel skins of factory door and wall panels. Although the mixed liquid ingredients are mobile, the rapid reaction and increase in viscosity mean that the placement of the material has to be facilitated by optimum mould orientation (in the case of the refrigerator), or by placing an extended nozzle of the mixhead into the cavity, then gradually withdrawing it as local filling is achieved (Figure 11.4).

The ease with which PU ingredients can be processed permits their application away from the controlled factory environment, using in-situ dispense equipment, such as on a building site for cavity wall-fill or storage tank insulation. A diesel generator runs high-pressure pumps and heaters which are used to dispense materials through an impingement mix-head; ingredients can be pumped a distance of up to 100 m. Mix-heads can be fitted

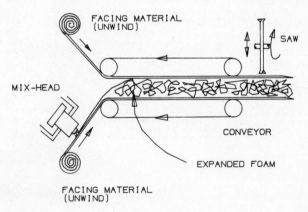

Figure 11.3 Diagram of the continuous production of rigid foam-filled panels

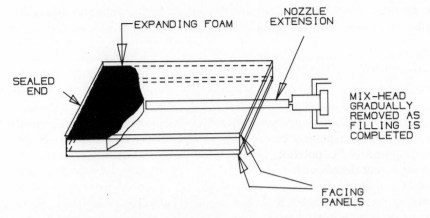

Figure 11.4 Diagram of batch production of rigid foam-filled panels

with nozzles for either spraying (to cover a large surface area) or pouring (for cavity fill). Ingredients are pre-heated to ease pumping, overcome low ambient temperatures and give very high reaction rates (less than 10 seconds), preventing slump on vertical walls. Damp conditions have to be avoided. Thick elastomeric coatings or sheet are used to protect the cured PU foam from direct sunlight.

Moulded high-density rigid foamed PU products ($200-750\,\mathrm{kg\,m^{-3}}$), are finding increasing use, for example in energy-absorbing blocks for car

Figure 11.5 A chest made up of decorative 'antique' panels based on barrier release coated rigid PU foam. By permission of Oakleaf Reproductions

bumpers, decorative panelling, or soles for clogs. Because of the initial low viscosity and wetting properties of the mixed ingredients, a PU will accurately reproduce the detail of a mould surface. This is particularly exploited in flexible mould techniques. Usually a wooden carving (the *master*) is made, and a silicone rubber *pull* cast from it, which is used as the liner of an epoxide mould. To increase the life of the silicone mould (up to 100 mouldings), an *in-mould barrier release* surface coating has to be used. The coating also means that a component out of the mould will process in the same way as a primer-coated piece of timber: it can be finished using traditional wood-staining and lacquering techniques. The finished product has all the appearance of the original master. Use of flexible silicone or PU elastomeric liners permits a high degree of three-dimensional relief in the product. Kitchen cabinet doors, decorative panels and beams, and ornate picture frames are made by this technique (Figure 11.5).

11.3.3 *Reaction injection moulded polyurethane products*

The advent of high-pressure metering and impingement mixing of PU ingredients through *self-cleaning* mix-heads has provided a process method which allows PUs to compete with high-quality thermoplastic injection moulding, for example for large automotive panels and electronic component housings. Although some products can be made using high-pressure open-pour moulding techniques, reaction injection moulding (RIM) involves closed moulding processing of flexible or rigid self-skinned foams, or high-modulus elastomeric products. Modified RIM equipment is used to process reinforcing fibre-filled PU/polyurea systems (RRIM).

RIM cycle times range from 45 seconds to 10 minutes, although the application and drying time of release agent may take up to 60 seconds. In the mass production of smaller mouldings, it is not prohibitively expensive to employ a number of identical moulding tools (based on epoxide liners in aluminium for example) loaded on to a carousel moulding unit. The carousel can hold up to 50 moulds; these progress past a lightweight RIM mix-head, which can be readily attached to and detached from each mould for filling. This allows the manufacturer a high output, although unit cycle time may be about 10 minutes. In PU-RIM systems having longer mould cycles, the high-pressure dispense unit can be used to feed several mix-heads bolted on to separate moulds in their own clamping presses, via ring-main/manifold units.

Mixed ingredients leave the mix-head nozzle at velocities up to $2\,\mathrm{m\,s^{-1}}$, and gating into the mould cavity must be carefully designed to prevent turbulent flow with the associated problem of air entrapment. As PU systems may react within a few seconds, filling of larger mould cavities becomes particularly critical to prevent underfilling or weld lines, which will give structural weakness in the product. Liquid PU ingredients will readily flow into the mould cavity, but the low pressure involved means that mould orientation, the filling point and split-line locations must be chosen with care, to ensure that air is fully expelled by the moving front of liquid ingredients entering the cavity. Possible moulding faults include underfilling, air-traps and density gradients in cellular products (in vertical or inclined sections).

PU-RIM systems are normally microcellular, high-modulus elastomers, based on polyether polyols – 'pure' MDI systems, although better high-temperature performance is achieved using polyamine – polyether polyol systems. Foam formation and self-skinning is promoted by trichlorofluoromethane (TCFM) or methylene chloride blowing agents. The ingredients injected into the mould cavity react and exotherm, boiling the TCFM and starting the foam formation. A little of the TCFM will escape into the air space of the mould cavity, rapidly condensing on the walls of the mould under pressure. The combination of the foam collapsing as it expands into the mould walls, due to the surfactant action of the liquid TCFM and overpacking of the foam in the mould cavity, produces a high-density skin on the surface of the moulding. Automotive steering wheels and head restraints are produced in this way.

Both products are made in a similar manner. Release agent is applied to the clean mould surface and allowed to dry. To overcome the problems of poor UV stability and to reduce production costs, a surface coating is sprayed on to the mould surface and allowed to dry. Mould temperature is 40–90°C to speed

Figure 11.6 Production of a car steering wheel moulding using modified RIM equipment. By permission of Sheller Clifford

paint drying and, subsequently, the foam reaction (Figure 11.6). A metal insert, intended to give rigidity to the product, is placed in the mould, which is then closed. A high-pressure mixing-head is clamped to the inlet port of the mould and a shot of mixed ingredients fired into the cavity. The head is swung away and the expanding foam is allowed to just spill out of the mould before its inlet is plugged. The mould is heated to speed the reaction, either by internal water channelling or by transferring it to an oven, in the case of a number of small moulds on a carousel line. The product can be demoulded in a few minutes, and deflashing removes a thin line of oversprayed paint. The *in-mould* PU paint appears to give near-perfect adhesion with the substrate. Shoe-soles are made from polyester – 'pure' MDI systems using similar techniques: again, dual-density foams can be made in a one-step process by an index change of the reactants.

(i) *Self-skinned semi-flexible polyurethanes.*   These give a soft and comfortable feel, with excellent abrasion and stain resistance. The foam will act as a shock absorber, for instance if a car driver is thrown against a steering wheel or head restraint. The poor resistance to UV exposure and the limited range of colours of PU has led to the successful use of barrier coats based on non-aromatic isocyanate-based-PU surface coatings. The accuracy with which PU reproduces the detail of the mould surface allows the stylist to use subtle textures. PU currently dominates the steering-wheel market (approximately 80% in Europe); polypropylene has the only other significant share. Although subject to fashion changes, the PU head-restraints are used in many down-market vehicles. Shot weights of ingredients will be under 750 g, with less than 5% wastage (from runners, etc.). Integral-skin components made in one production step have replaced components fabricated from stitched fabric or rotational moulded plasticized PVC, reinforced with a steel frame and filled with PU foam in a second process operation (Figure 11.7).

(ii) *Self-skinned rigid polyurethanes.*   These are used for office and kitchen furniture and housings for electronic components such as television cabinets in the range 4–15 kg moulded weight, and are made by using dedicated mix-head/mould combinations. Many of these foamed products have to be surface-coated. From the mould, the product is allowed to post-cure at room temperature before solvent-degreasing to remove release agent from its surface. Degreasing is aided by sanding using emery cloth, which will also eliminate the flash-line scar while creating fine scratches in the surface to enhance paint adhesion. Any small holes in the moulding's surface have to be filled at this stage using unsaturated polyester or cellulosic putty. The moulding is normally primer-coated, then lightly resanded before a topcoat paint is spray-applied. Generally the post-moulding operations are the time-consuming and labour-intensive steps in the production; because of this, rigid

Figure 11.7 Range of in-mould coated, semi-rigid, self-skinned PU foam mouldings. By permission of Marley Foam, Sheller Clifford and Bridgtown Industries

Figure 11.8 Preparation of a chair moulding based on rigid PU-RIM for post-mould painting

integral-skinned PUs face severe competition from self-coloured foamed thermoplastics like ABS and polycarbonate. Compared with non-foamed plastics such as modified PP, foamed PU in particular offers the designer the opportunity to use thicker sections (10–20 mm) giving 'body' and elegance to the shape of a piece of furniture (Figure 11.8).

(iii) *Solid and micro-cellular elastomeric polyurethane RIM products.* These are used largely in the production of exterior automotive components: wheel arches, bumpers, panels, and many add-on units employed to enhance the appearance and cost of up-market versions of cars. Legislation to improve fuel consumption by weight reduction and to reduce collision impact damage has led to the acceptance of plastics as car bumpers, replacing chromed steel. In the USA, the original 5 mph requirements for damage control resulted in the use of high-modulus PU elastomers, which would buckle under the impact load then fully recover, even at speeds up to 10 mph. Vehicle stylists realized that plastic bumpers could be designed to provide large complex shapes with a greater number of functions than the traditional metal components, for example complete (soft) front and rear ends for cars. In Europe, the 5 kph (3 mph) requirements subsequently introduced by the EEC opened the door to engineering thermoplastics in preference to PUs (exceptions included the Jaguar XJS and MGB GT bumpers).

Some automotive companies such as British Leyland (now Austin Rover) prefer complete on-line painting for exterior panels and bumper assemblies, meaning that any plastics component will have to withstand the same paint stoving conditions (150°C or higher, for up to 30 minute) that sheet steel undergoes. Under these regimes, PUs will suffer heat distortion, while modified polycarbonates will become stress-crazed by the paint solvent. The alternative is to pigment and stabilize the moulding in a neutral colour: black for PUs, light grey or brown for polycarbonate—polyester as used on the European Ford Sierra. Few plastics can take such stoving conditions; these include modified poly(butylene terephthalate), unsaturated polyester SMC, and glass-reinforced polyurethane/polyurea and polyureas. Short-strand glass fibre (up to 20% loading by weight) in the PU/polyurea system increases modulus, reduces coefficient of thermal expansion (when installed with metal panels, thermal expansion and contraction is less likely to show either as distortion around the fixing points or as large gaps between the panels, respectively), and reduces heat sag and creep in service (Table 11.1). In the USA, there is relatively large volume use of reinforced PU–polyurea, as in the Pontiac Fiero's body panels, but a radical change in vehicle assembly methods has had to be adopted (Figure 11.9). In Europe these materials will be found only on low-volume-number vehicles, e.g. Porsche (soft rear ends), Reliant Scimitar SS1 (body panels), Rolls Royce Silver Spirit (front spoiler), although it may be speculated that in mass vehicle production there will be wider acceptance as faster PU or polyurea systems become available.

Figure 11.9 Pontiac Fiero made up of exterior plastics components fixed on to a load-bearing steel skeleton, including PU-RIM side panels, glass reinforced PU-RIM door panels and polyester SMC bonnet and roof panels. By permission of Bayer (UK)/UT

11.3.4 *Elastomeric and resinous thermosetting polyurethanes*

Good abrasion and chemical resistance in many environments, combined with ease of application at room temperature, has led to wide use of PU elastomers. Products range from covers for printing-press rollers, abrasion-resistant lining materials, to low-speed tyre filling and vehicle floor mats. Synthesis can be either by *one-shot* (polyol reacted with isocyanate) or *two-shot* process (a prepolymer is first made, normally with isocyanate-tipping, which can later be reacted with a curing agent, e.g. diols as chain extenders or polyamines or triols as cross-linking agents), or with atmospheric moisture if applied as a thin film. Diamine-cured prepolymers give excellent products; an example is the sound-proofing covers for pneumatic drills used in mining and quarrying where good environmental and abrasion resistance is a prerequisite. Polyether–MDI systems can be sprayed in factory or in-situ applications, to provide abrasion- and corrosion-resistant coatings (in quarry conveyor buckets, hovercraft propeller blades, water-inlet ducts for power stations and in many civil constructions such as oil rigs; in these applications the PU may need to be applied in 'films' up to 50 mm thick—Figure 11.10). Energy-absorbing PU elastomers (e.g. Sorbathane™) are used to line motor-cycle crash-helmets; these are based on polyether polyol–MDI systems where either plasticizers or excess polyol (low isocyanate index) are used.

Figure 11.10 In-situ spray application of a 100% solids PU elastomer to the interior of a pipe. By permission of Lithgow Saekaphen Ltd

The resinous or rigid PU systems lend themselves to room-temperature mixing and dispensing, often by hand. High loadings of selected fillers are used as cheapeners, to improve electrical properties and give dimensional stability in mouldings–hence their use in the encapsulation of electronic components (such as in starters for street lighting), as insulators for high-tension power lines, on-site applied sealants for cable splice boxes, and for self-levelling flooring systems. PUs compete with epoxide resins in this field.

11.3.5 Thermoplastic polyurethanes

In comparison to other types of PUs, thermoplastic polyurethanes (TPPU) are of limited availability and end-use, although they were amongst the first PUs to be developed in the early 1940s, as competition to polyamide fibres. Again they are exploited for their abrasion resistance, over a range of hardnesses from 50 Shore A to 80 Shore D. Basic thermoplastic injection-moulding and extrusion equipment can be used to process TPPU, although the polymer granules should be dried before melt processing, to prevent degradation. Uses include the exotic—TPPU replaces goose feathers as flights for archery arrows, because the snap-back characteristics are similar. It has also been used for combined sleeping policeman/cable protectors for geophone cables used in built-up areas. TPPUs can be blended into plasticized PVC and elastomer formulations, to improve stiffness and modulus without loss of low-temperature properties, for instance in shoe soles. Recently a range of alicylic isocyanate/caprolactone polyester TPPUs has been developed for UV-resistant, water-white safety materials such as faceshields. These have superior abrasion resistance to polycarbonate, more commonly found in this application.

11.3.6 Polyurethanes as surface coatings, adhesives and sealants

A wide range of starting materials can be used to form PU with virtually any selected properties. This is nowhere better exploited than in the surface coatings and adhesives industries. The chemistry and surface-tension properties of PUs allow wetting and excellent physical interfacial bonding with a wide range of substrate materials, and this is enhanced either by allowing PU

Figure 11.11 A two-pack, low stove PU vehicle paint system, used as an exterior finish for the Panther Kallista

reactants to react directly on the substrate's surface, where isocyanate will tend to form strong covalent bonds, or by dispersing reactants in suitable solvent carriers.

The abrasion characteristics of PUs are again exploited in surface coatings, along with resistance to chemicals, ease of application, and low-temperature drying and curing. Uses include urethane alkyds for house paint: one- and two-pack paint systems (as motor vehicle finishing and refinishing paints, Figure 11.11), varnishes (e.g. for marine timber and electrical wire lacquers) or inks. Isocyanate-tipped PU prepolymer two-pack systems are popular in industrial applications, curing by reaction either with atmospheric moisture or with suitable curing agents. UV-curable PU/acrylics are used as paper finishes and cabinet door lacquers.

Adhesives can be based on either reactive or fully reacted PU systems. Large quantities of PU are used in contact adhesives in the construction industry. PU sealants normally rely on moisture curing, for instance sealants used for wooden frame constructions in domestic housing.

Further reading

Hepburn, C., *Polyurethane Elastomers*, Elsevier-Applied Science, Amsterdam (1982).
Woods, G., *Flexible Polyurethane Foams*, Elsevier-Applied Science, Amsterdam (1982).
Becker, W., *Reaction Injection Moulding*, Van Nostrand Reinhold, New York (1979).
—, *Polyurethanes*, Bayer publ. (PU50025e), Bayer AG, Frankfurt (1979).
Woods, G., *The ICI Polyurethanes Book*, Wiley, Chichester (1987).

Index